Naked Eye

ASTRONOMY

How to Read
the Heavens

By

Zahir K. Dhalla

Other works by the author

at Amazon.com

- "Poetry: The Magic of Few Words"

- "My F-word Plan: How I Routinely Maintain Low Weight & Good Health"

- "Nine Ginans of Nine Ismaili Pirs: A Brief History of Khoja Ismailis"

- "Learn Good SWAHILI: Step by Step A Language Textbook"

- "The Willowdale Jamat Khana Story"

- "My Tanga Days: 1950s & 60s"

- "Learn Urdu: اردو – Read, Write Speak"

Naked Eye ASTRONOMY

How to Read the Heavens

1st print edition published:

February, 2019

By Kitabu Publications, Toronto

(A division of 1304266 Ontario Incorporated)

kitabupublications@gmail.com

ISBN 13: 9781091243286

Cover Design: by the author

All proceeds of this book will go to needy school children in Tanga, Tanzania (where I was born, grew up and finished high school).

To

All

Who Gaze Heavenward

And Can't Help Wondering,

In Awe!

TABLE OF CONTENTS

TABLE OF FIGURES

1. Introduction

Twinkle twinkle little star

How I wonder what you are.

Mozart's famous ditty.

Ever since 1972 when I first measured stars, from the roof top of the engineering building, for my astronomy course at the University of Nairobi, Kenya, I have always gazed heavenward whenever an opportunity presented, and the deep sky never ceased to amaze me. Simply being able to imagine the seemingly 2D (two dimensional) starry sky as the true 3D space that it actually is, is a thrill, to say the least. And on top of that, being able to identify stars and their relative positions added to that thrill. I tell you, when you can point to a star or a planet and say "That's Pollux, one of the twin stars of Zodiac constellation Gemini!", for example, it is most satisfying.

This visualization and identification was not readily acquired by me. It took years and decades (mainly because there was no Internet to speak of in the 20th century), as a part-time hobby, but has been most rewarding. My goal in this book is to share all this with those of you who have looked up and wondered what are all those stars out there and their inter-relationship. For those of you who already have this knowledge then perhaps this book will serve as another validation.

In this 21st century, we can thank our lucky stars (!) for the help we are getting from Internet and computer technology, yet even with a computer-generated sky view, it still is a challenge to recognize what is what, and the when and the how. One has to at least know the local Latitude and Longitude, how to orient towards North or South, etc, but in addition to these minimums, one has to know the various coordinate systems in use and their relationship to Sidereal Time, blackout zones, etc and then make observations in your sky based on these references. These references are also most necessary when surfing the web for useful info on astronomy topics, for being able to understand the jargon. This book's goal is to provide a solid foundation, in one place as compared to scattered across a vast array of references, from which to then take off. After that, only the sky's the limit!

Accuracy

All observations herein are by eye, measured using fist-n-thumb. The best accuracy using these techniques is about 1°. The hour equivalent of this is about 0.1h (1° ÷ 15, rounded to one decimal point). Accordingly, degrees herein are in whole degrees and hours to one decimal point. This rounding to whole degrees and to one decimal point of an hour results in a maximum error of ±0.5° or ±0.05h (= ±0.75°). For our purposes of observing stars, constellations and planets, either to identify them or to locate them, this accuracy is more than sufficient. [Hours herein are given in decimal points and not as hh:mm, as the latter always has to be converted to decimal mode anyway in order to use them in calculations :-)]

Apparent Rotation of the Sky

One of the very visible features of the night sky is that, because the earth *rotates around its North-South axis* (imagine a spinning globe of a geography class) *every 24 hours*, so our view of the night sky appears to rotate. Thus, what you see straight up overhead (called the **Zenith**) at say 9 pm will be different than say at midnight.

While we are on the subject of earth's movement, let's go through them all - and then dispense with them altogether for they do not matter to naked eye astronomy, as *rotation* does. This is only background info covering all of earth's movements. These movements are caused by the gravitational effects of the various heavenly bodies – the sun, giant planets like Jupiter, the nearby moon.

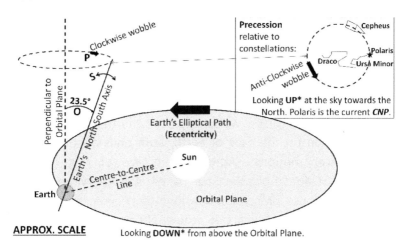

1 Earth's Other Motions

P = **Precession** i.e. the Earth's axial tilt wobbles in a circle

O = **Obliquity** = Earth's axial tilt

S = Swings of the Earth's axial tilt

*Notice in the **UP** inset the direction is anti-clockwise, while in the **DOWN** main diagram it is clockwise. This is analogous to viewing a clock from behind it: the direction reverses.

Earth's Revolution around the Sun

[Period: 365.25 days] The earth completes one revolution around the sun but for our purposes this movement does *not* alter the positions of stars on the fictitious **Celestial Sphere** on which they are assumed to be located (they are actually at widely varying distances from us but all being so far away we can conveniently assume that they are located on the surface of one infinitely huge sphere). At the scale of this **CS**, the whole revolution path of the earth is a mere tiny, tiny dot. An observer on the surface of the **CS** can not resolve the solar system into its components and therefore would not discern their movement around the sun, and so, vice versa, we can not discern the effect of our revolution when projected on to the **CS**.

Earth's Precession

[Period: 25K+ years] Another movement of the earth is its wobble, like how a spinning top wobbles whose 'north pole' makes a circular motion (in the case of earth, both ends of its spin axis make circular motions). Although the size of this circle is significant* it does not affect our naked eye observations for a long time because precession is a long-term cycle, so that only after a few hundred years it would affect our observations, 'would' because the star coordinate system in use is updated to allow for this. [*For example, currently the North end of earth's spin axis points approximately to Polaris in Ursa Minor constellation. Over the next thousands of years, it will point to different constellations e.g. half way through the current wobble cycle it will point to constellation Lyra which is well over a distance of a quarter of the sky from Polaris (the total sky distance being horizon-to-zenith-to-opposite-horizon). Also, an interesting observation is that although it wobbles, it does not change Earth's axial tilt, just like a spinning top maintains its tilt albeit eventually it slows down and the tilt increases – Earth's spin though is steady.] There are two other motions of the earth's spin axis, measured in metres at the poles. These are *nutation* and *polar motion*. The former is caused by the same forces as those causing precession, except precession is steady, large and long-term while nutation is the opposite. Polar motion is a change in the location of the spinning axis with respect to earth, caused by the redistribution of earth's mass (e.g. by melting ice sheets). This is to be contrasted with precession and nutation which are not with respect to earth itself but are observable external to earth.

Earth's Axial Tilt [=Obliquity]

[Period: 40K± years] The earth's spinning axis is *not perpendicular* to its plane of its motion around the sun. [It IS a *flat* plane, meaning the earth does not bob up and down along its path around the sun.] It is titled from the perpendicular by **23.5°**. But this tilt increases and decreases by a couple of degrees (like the top of a ticking metronome's pendulum). Again this does not affect our naked eye observations because its long 'ticking' cycle is tens of thousands of years.

Earth's Eccentricity

[Period: 100K± years] Earth's path around the sun is not exactly a circle, it's an ellipse, a squashed circle so to speak. How much it deviates from a perfect circle is called eccentricity. This eccentricity increases and decreases (slightly), from a near perfect circle to a slightly less one. Again, we can ignore this for our purpose.

Stellar Motion

In addition to the above earthly motions, the stars are also on the move. The stars of our Milky Way galaxy orbit its centre at high speed e.g. our star – the Sun – and our solar system fly around the galactic centre at half a million miles an hour, taking a quarter of a billion years to make a complete circuit. These motions do not alter our view of the stars relative to each other except over eons.]

Net Effect

In a nutshell, the good news, for our purposes, is that we only need to deal with the net, practical effect of all these motions, which is that as the clock ticks away the night sky rotates in a simple, consistent way, day after day after day... Imagine a geography class spinning globe: as it spins, the continents and oceans on its surface spin, but their positions relative to each other as well as to the globe remain fixed. And so it is with the sky: the heavenly bodies rotate from East to West (like the Sun), while their positions relative to each other remain fixed.

Sidereal Time

The effect of earth's rotation on what we see in the sky is expressed by use of an appropriate time clock. Our regular clocks are sun-based, that is the average time from one sunrise to the next is 24 hours. However, if you were to measure the time from one star's rise above the horizon to its next, it will not be 24 hours but about 4 minutes short of it. Our 24-hour clock is termed solar time, while the shorter star day is called *Sidereal* (star) *Time*. The cause of the difference between the two times is Earth's movement around the Sun. Earth, as it rotates around its axis, it also revolves around the Sun. Between one sunrise and the next, the Sun 'falls behind' slightly due to Earth having moved forward in its revolution around the Sun. This results in the Sun rising approximately four minutes late every day. [This is diagrammatically shown in figure **70 Sidereal Day** under the *Sidereal Time* entry in the **Glossary**.] What does all this mean to our viewing the night sky? Say at 9 pm (based on our usual solar time) you look straight up (called the *Zenith*) and make note of a star there. The next day that star will be at *Zenith* earlier, at about 8:56 pm (solar time), because it keeps to star (*Sidereal*) *Time*. Astronomers use *Sidereal Time* for ease of calculating star positions, as it makes for a consistent / routine 'star day'. The good news for us is that a website showing star positions for a given location and time will quite likely let the user type in the usual solar time, thus no worries about **Sidereal Time**. But even if input is required in **Sidereal Time**, conversion calculators are available on the Internet and as smart phone apps :-)

TIP: The various **URL links** given in this book can change as their owners make changes, however its host name is usually not changed (unless there has been a major reorganization under a different host name, but even in that case there usually is a redirection from the old host to the new) e.g. you are given a URL link http://www.acme.com/Paris-metro/ and you find it does not exist. You can then try the host name http://www.acme.com/ (which usually would be found) and then find a "Paris Metro" link. Failing which, a search engine on the Internet would have to be used, to search "Paris Metro".

How to Use this Book

Like any textbook, the first step of course is to familiarize yourself with its organization by browsing the table of contents. Next quest is how the book is laid out. This is achieved by a quick browse of the contents of each chapter. This is followed by a slow reading of the material from start to end but not necessarily everything will be grasped. From here on, the book is read selectively with the objective of grasping what is being described, and involving much cross-referencing* of material in different chapters. [*Cross-referencing includes looking up technical terms in the **Glossary**, which elsewhere in the text are italicized and bolded.] This last process is not a one-time thing but rather an on-going one, because alongside learning and grasping the descriptions / explanations, is applying these to field work, which will require a back-n-forth between the book and the field experience. [After all, it took astronomers of yesteryears, millennia to figure out the night sky's geometrical behaviour. Thus, if things are not grasped right away, it behooves us to keep in mind the long time it took to bring the field of astronomy to the present. The good news is that we do not have to *figure* anything out. That's been done for us. For us then, it is to *grasp* it. And grasp it, we can – over time.] In general, this approach of reading-observing-reading-and-so-on pays handsome dividends. I can attest to that from my own experience. And doing this over sufficient time, your familiarity with the sky will mature to the point where you can refer to it as part of your 'neighbourhood'. Most satisfying, I tell you.

Good luck, and have lots of fun.

2. What is Out There?

We will explore this in a top-down manner, that is outer to inner, as follows:

3 A Tour of the Universe

4 A Tour of our Galaxy

5 A Tour of our Night Sky

The universe is a set of galaxies (of which ours is the Milky Way galaxy), which in turn are sets of stellar systems, one of which is our solar system consisting of the sun and all the bodies that revolve around it. The sizes of all these sets and sub-sets is as follows:

- A ray of light would take about a 100 billion years to traverse the universe.
- The ray of light would have taken 'only' about a $1/10^{th}$ of one million years to cross our Milky Way galaxy.
- That ray of light would have taken less than a couple of years to cut through our solar system.
- Lastly, the ray would have taken less than 10 minutes to travel from the sun to the earth.

Thus, the universe is about a 1000 tera times bigger than the Milky Way galaxy which in turn is about 50,000 times bigger than our solar system. Another way to appreciate these sizes is to consider the Dog Star (Sirius), the brightest* star in the sky. It's about 8.5 light years away from Earth. When you look at it you're seeing its state 8.5 years ago, meaning its light that you are now viewing left it 8.5 years ago! [*Brightness in astronomy is given as *Magnitude*.]

So that's the '**What'** of the sets of the various entities of the universe. But there is of course the question that many have asked: '**Who** is out there?' The answer apparently is simple: no one – yet. This answer is based on the simple explanation that there is no evidence (communication really) on the existence of life outside our Earth. But an argument can be made that our communication science is not necessarily at the same level of sophistication as that of potential 'civilization(s)' out there. That is not to say that scientists are not trying to find out given our level of sophistication. One of these efforts is locating planets outside our solar system (called exoplanets). Although these can not be visually observed, their presence can be detected by indirect means, being basically measurement of stellar phenomena that can be explained by the presence of bodies around stars.

3. A Tour of the Universe

The **universe** is a vast, vast space, containing many, many billions of **galaxies**. Galaxies are collections of many, many billions of stars. Galaxies themselves are clustered, and these **clusters** are further clustered into **superclusters**. Despite this huge population of stars, they make up only a tiny portion of the volume of the universe, the rest of it being mostly a void!

Our **Solar System** is made up of one star, the **Sun**. It belongs to the **Milky Way Galaxy** of stars, which in turn belongs to the **Local Group Cluster** of galaxies, which in turn is part of the **Local Supercluster**, situated somewhere in the **Universe**.

All visible stars in our sky **belong only** to the **Milky Way** galaxy, and even these are only the ones that are near to us, meaning only a fraction of our galaxy is visible to us. Not only are most of the stars of our galaxy not visible to us but so too are all the stars outside it in the vast universe. [A few nearby galaxies are visible to us as fuzzy whitish smudges, but not their stars: The **Small Magellanic Cloud** (SMC) near the border between constellations Hydrus and Tucana, is only two Milky Way diameters away from us and less than 10% as wide as our galaxy. The **Large Magellanic Cloud** near the border between constellations Dorado and Mensa, is closer to us and twice the size of the SMC. **Andromeda Galaxy** in constellation Andromeda, is more than ten times far away as the SMC, but being twice the size of our galaxy, it is visible.]. The moral of the story is that we can only see the relatively close*, bright stars of our galaxy and perhaps the odd galaxy (as a single object) beyond. The rest of the vast, vast universe is invisible to us. [*Though relatively close, the visible stars are nevertheless very far away e.g. the earliest space probes Voyager I and II have now left our star system (our solar system) 40+ years after their launch. The two Voyagers will pass by nearby stars in about

40,000 years! Furthermore, by 'pass by' is meant within 1 to 2 light years! What this also illustrates is that although there are billions and billions of stars in our galaxy, most of the galactic volume is made up of empty space sprinkled with dust and other particles, so that if one were to shoot a theoretical, powerful shotgun into space, the chances of a shot bumping into a heavenly body is almost nil!]

4. A Tour of our Galaxy, the Milky Way

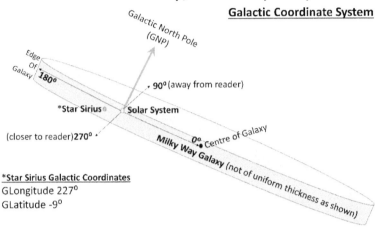

Galactic Coordinate System

2 Galactic Coordinate System

In the above diagram, the Milky Way (our home galaxy) is represented by a thinnish grey disc. Imagine the grey represents some liquid. Now imagine that a whole bunch of very small beads (not shown in the figure), of different sizes and brightness are randomly suspended in the water - that is, some float near the bottom of the disc, some near the top, the rest randomly in between. Each bead represents a star and collectively they represent the Milky Way galaxy, while the 'liquid' of our model represents the inter-stellar void sprinkled with dust and particles. Our home star, the Sun (labelled Solar System in the diagram), would be a tiny little bead about half way between the centre of the disc and its edge, and about half way between the top and the bottom of the disc.

The Sun bead should be treated as our whole solar system (being the Sun and all its orbiting bodies) since at this scale the planetary bodies cannot be separately resolved from the Sun. Then, what we see in our sky from Earth is how the Sun bead views the other star beads. Thus, the Earth within this bead sees star beads all around it, since it is suspended halfway up from the bottom of the disc. But only the closer and brighter, big stars are visible to the naked eye. How close? Less than 15,000 light years or so (being the distance covered in 15,000 years, travelling at the speed of light, which equals to 150,000 trillion kilometres). The average radius of the Milky Way galaxy is about 50,000 light years to use a nice round number. Thus, we are able to see stars up to about two-thirds of the way towards the centre of the galaxy (being the middle of the disc in our model) and, looking the opposite way, again up to two-thirds of the way towards the edge of the galaxy. As for either side of this radial line (centre-Sun-edge), we can see up to 15,000 light years away. The remaining dimension, the thickness of the disc, is only about 1,000+ light years thick: thus, we can see all the bright, big stars all the way to the top and all the way to the bottom of the disc. These are the furthest distances we can possibly see, but the ten brightest stars seen from Earth are less than 900 light years away, and six of these are less than 50 light years away, the brightest of them – Sirius, the Dog Star – being less than 10 light years away.

Beyond the 15,000 light years visibility, the rest of the stars of the galaxy (in all directions, towards the centre and towards the edges) are only seen as a thin band of greyish white glow. This glowing band has been referred to, unfortunately, as the **Milky Way** and yet all the other stars we see in the night sky outside this band are also physically in the same Milky Way galaxy as the band! Hopefully our disc model lets you put the visible stars and the glowing band in proper perspective :-) Note that the Milky Way band is not visible in urban areas nor even in less lit-up places. This is because of light 'pollution' whose direct and diffused glow interferes with the dimly-lit stars. You have to go out into uninhabited land to see it. A pier jutting out into a secluded lake would be a good candidate for observing the Milky Way band. [I did something like that on the North-East shores of Lake Simcoe, Ontario, at Brechin Beach, which is relatively secluded. Secluded enough it was that I was able to trace a faint greyish band through the sky. Which brings to mind my growing up days (1950s, 60s) on the coast of East Africa. If you asked me 'What is the Milky Way?' I would readily point to it in the sky, it was that obviously visible. Ironically, 45 years later when I returned to my home town for the first time after leaving, I couldn't see it :-(Presumably, light 'pollution' is now global (just like smog is). It's either that or my eyes are no longer young – or both! [There is also another attenuation, besides light pollution: atmospheric attenuation, due to the contents of the air mass above us viz. gas, vapour, dust… The depth of this air mass being smallest above us and widest at horizon level – a 1:40 ratio – the attenuation is the least above us, most at the horizon.]

Galactic Coordinate System

[Refer to the above figure.] We can imagine reference lines relative to our disc model which can serve as the axes of a coordinate system for the galaxy. The line from our Sun bead to the centre of the disc is our 0° line. The line from our Sun bead to the nearest edge of the disk, being opposite to 0°, is our 180° line. These two lines make up a single radial line through the Sun bead. Starting from the 0° line, the other lines are established viz. 90°, 180° and 270° in an anti-clockwise order when looking "down" from an observation point "above" the galaxy. ["down"/"up" and "above"/"below" are arbitrary in our model as from a universe / space perspective there is no "up" nor "down".] These directions are **Galactic Longitudes**. **Galactic Latitudes** are angles above and below the galactic plane (the central plane of the disc slicing through our Sun).

For our purposes, this galactic coordinate system can be ignored as it is not used in everyday astronomy. However, it is useful for establishing where a star is in the context of our galaxy, specifically when the Milky Way band can not be seen in urban areas (an even where it is visible, we would need to know where one of its four cardinal points is). For example, **Sirius** the Dog Star, has a galactic Latitude of -9°, meaning that it is nine degrees (9/10 of a held-out fist) below the Milky Way band. But we can't tell what the tilt of the band at this location is. However, **Procyon**, the 8th brightest has a galactic Latitude of +13°, meaning that it is thirteen degrees (a fist and a third) above the Milky Way band. *Now*, we can tell what the band's tilt is between these two very bright stars (they belong to neighbouring and similarly named constellations, **Canis Major** and **Canis Minor** – Big Dog and Little Dog), because we know how far off each star the band is. So, even if we can't see the galaxy band, we know precisely where it is. Incidentally, the thickness of the band is only a couple of degrees (a held-out thumb width would be a good average). In addition, the galactic longitudes of these two stars, Sirius and Procyon, are 227° and 214°, respectively, about a fist and a third apart but more importantly, that many degrees away from the centre of the galaxy (0°). [To obtain galactic coordinates, use a good website that provides them e.g. see **Appendix D: Celestial & Galactic Coordinates of Stars**. For your convenience, **Appendix L: Information Tables** contains information that includes Galactic Coordinates.]

Other markers are the four constellations to which the four cardinal directions point to, and to the Galactic North/South Poles:

0° is towards **Sagittarius** constellation

90° is towards **Cygnus** constellation

180° is towards **Taurus** constellation

270° is towards **Vela** constellation

Galactic North Pole (*GNP*) (right angles to galactic plane) is towards constellation **Coma Berenices**.

Galactic South Pole (*GSP*) (also right angles to galactic plane) is towards constellation **Sculptor**.

These Galactic Longitudes and Latitudes are to be compared with Celestial Coordinates of the stars in our sky which are specified as *Right Ascensions* and *Declinations*. This latter system is the one which is commonly used in astronomy.

5. A Tour of our Night Sky, Part 1

This is it! This and the chapters that follow form the main theme of this book!

Our Milky Way Stars

Image our head being the Earth inside the Sun bead in our previously seen galaxy disc model (Figure **2 Galactic Coordinate System**). (1) Now look up to the top of the disc, (2) then to its bottom and (3) then scan horizontally 360°. We should see stars in all directions but their densities in these three directions will be different. There are fewer stars in directions (1) and (2), as compared to (3). The denser 360° view of (3), the **Milky Way** band, is explored further below. [This density difference is only visible in unlit areas, with the result that there is no apparent density difference in areas where we live. Interestingly though, there are NO named stars (read that as, bright stars) at the **GNP** nor the **GSP** within a radius of about 10°. There are many stars in those regions, but only dim ones. What is curious though, is that the same is true of the **CNP** and the **CSP**. Curious because these points are not in the same directions as the **GNP** and the **GSP**. Far from it. I can only surmise that there are high density interstellar dust clouds in those directions, blocking our view of stars. Still, very curious!]

Measurement by Fist-n-thumb

How big is the sky? From any point on the **horizon** to the overhead point (the **Zenith**), then continuing on to its opposite point on the horizon is an arc of 180°. If you fully extend your hand and make a fist, it will cover approximately an arc of 10° in the sky i.e. about 18 of them can cover the above 180°, or 9 from horizon to **Zenith**. A thumb width covers very roughly 2°. Human fists and thumbs not all being the same size, they need to be calibrated and adjusted for:

Calibration of fist: Fully extend your hand, make a fist and measure how many fists it takes to go from a point on the horizon to the **Zenith**, and as a check, continuing on to the point on the horizon opposite to the above point. My fist being small, I counted 12 ½ fists from horizon to **Zenith** (and my check to the other point on the horizon confirmed it). Ergo, my fist = 90 ÷ 12 ½ = 7.2°. As a further check, measure along the horizon from any point all the way around it coming back to the starting point. This would be 360°. Another way to calibrate, is to measure the fist width and the eye-to-end-of-extended-fist distance. Dividing the first number by the second gives you an angular measurement in radians which can be converted to degrees by multiplying by 57.3. For example, in my case, it came to (8 ÷ 64) X 57.3 = 7.2°, which is the same as my above measurement.

Tip: Since fist-n-thumb sizes are not the same for everyone, in this book they are conventionally set to be 10°-n-2° respectively. Once you have calibrated your fist-n-thumb sizes, you will have to apply a personal adjustment e.g. stars Vega and Deneb are about 2½ fists apart, meaning about 25° (2½ X 10) apart. So, my fist being only 7.2°, my fist measurement of Vega-Deneb will be about 3½.

Stars 'revolve'

The first thing to appreciate is the apparent rotational movement of the stars. It is apparent, not real, being so because it is the earth which rotates. It's as if the earth is scanning the sky. This apparent movement is slow e.g. star Kochab, the second brightest star of the Little Dipper asterism [this is the ladle/dipper shape of constellation Ursa Minor, at whose handle's tip is the star **Polaris** where approximately the **Celestial North Pole** (**CNP**) is located], rotates around the **CNP** but it takes almost 24 hours to make this apparent circuit. If, however, you take a long-exposure photo of this you will see a trace of its rotational path, as you would of the other stars but not of Polaris which is the centre of this rotation and so does not have a circle of rotation. Another way to appreciate the effect of this rotation on the night sky is that, two locations at the same Latitude but in different time zones, say five hours apart, would see the *same* night sky five hours before or after (depending which point is East or West of the other).

TIP: Earth's rotation is from the West to the East, which makes the Sun and the stars and the planets ALL appear to rise in the East and set in the West.

TIP: 'Circling' stars: since stars move in their fixed circular paths around Polaris, their **Declination**s (**Decl**s) tell us how far they are from Polaris regardless of date-time of observation. And if they are currently above the horizon (meaning, visible), this information helps us half-locate them (the other half-location being where they are on these circles of rotation). For example, star Vega (5th brightest overall) is at **Decl39°** (angle above the **Celestial Equator**), meaning it is on a circle at an arc distance of 51° (90 – 39) from Polaris. Where on that circle it is, is taken up later herein, along with **Decl** and other terms.]

Constellations

Visible stars are grouped into constellations. These are visual groupings, not spatial, in that by sight they appear to be clustered together but spatially, in depth, they are far apart. For example, the approximate size of constellation **Orion** (one of the most easily recognized constellations) is 25° by 30°, but the distances of its stars from earth range over hundreds of light years, one light year being 10 tera million kilometres.

The ancient astronomers imagined skeletal shapes by selectively joining the dots (i.e. stars in the neighbourhood). Some saw in these stick drawings, skeletal shapes (many incomplete) of mythical figures, others of animals, and so on. [The constellations toured herein are identified by their name and the stick figures they represent e.g. **Orion the Hunter**.] Today, there are 88 constellations as agreed upon by astronomers. Their imaginary skeletal shapes facilitate star identification. The problem of course is that the imaginary lines that make up these shapes are not actually there in the sky connecting the stars. Amateur astronomers, still learning their hobby, have to use drawings of constellations to try and match them to the sky. In time, these – like aide memoires – would become memorized, especially the ones that are relatively easy to recognize. [An additional problem with constellation stick-figures is that often they are not consistent e.g. constellation Virgo is represented differently in the literature out there, two extreme versions of it having no resemblance to each other. Typically though, the differences are in terms of an additional one or more connector lines.] Constellations, despite some of these shortcomings, are still useful for relative positioning and recognizing. They are like a jigsaw puzzle of sorts. Of sorts because, although they do not interlock like jigsaw puzzle pieces do (their boundaries do interlock, but not the stick figures), they nevertheless are positionally inter-related just like puzzle pieces. While jigsaw puzzle pieces are inter-locked by matching them to its given complete picture, constellation stick figures are inter-positioned by using our acquired knowledge, neighbourhood by neighbourhood. For example, constellations **Cygnus the Swan** and **Aquila the Eagle** are neighbours, these birds appearing to be flying towards each other, and **Pegasus the Winged Horse** looking to joining them in a head-to-head meeting, with **Lyra the Lyre** providing a befitting score for this heavenly get-together! Thus, recognition of one constellation leads to one or more of its neighbours. [More on this later.] One more thing, the stick figures do NOT incorporate ALL of a constellation's stars, certainly not the ones discovered in relatively modern times. Nevertheless, the constellation boundaries as defined by the IAU (International

Astronomical Union) fully cover the whole **CS** and thus every star (even the ones yet to be discovered) belongs to one of the 88 constellations.

Asteroids, Comets, Meteors, Nebulae, Novae, Pulsars, Quasars

Under right conditions (little light pollution, high altitude), these celestial phenomena can be observed. What are they?

Asteroids are tiny, odd-shaped planets, circularly orbiting the sun, collectively in the **Asteroid Belt**, which lies between the four inner small rocky planets (viz. Mercury, Venus, Earth, Mars, in that order from the Sun) and the four outer giant gaseous planets (viz. Jupiter, Saturn, Uranus, Neptune).

Comets are big-to-huge icy chunks that elliptically orbit the Sun, their orbital periods ranging from years to eons. They originate either from the **Kuiper Belt** or the **Oort Cloud**. The Kuiper Belt is beyond the furthest planet, Neptune, and from it emanate comets of shorter orbital periods. The Oort Cloud is beyond the solar system, i.e. in interstellar space, and is home to comets of longer orbital periods. [Twin space probes Voyager 1 and 2 are headed for the Oort Cloud and will make a long, long journey through it, lasting tens of thousands of years.] It has been theorized that a comet originating in the Kuiper Belt, can move on to the Oort Cloud, as has been suggested for the **Hale-Bopp Comet**, seen in the 1990s (but you will have to wait a couple of thousand years to see it again)! On the other hand, **Comet Halley** is seen every ¾ century, the last time being 1986. As a comet comes closer to the Sun, it becomes warmer and emits some of its material as gasses called **coma**, which can lead to a very long **tail**, some visible to the naked eye. Some big comets are larger than Earth!

Meteoroids are small-to-tiny celestial debris e.g. left-overs of a dead comet or asteroid, which upon entering Earth's atmosphere, begin to burn and thus become visible. Visible meteoroids are called **meteors**, and any that survive the burning, fall to the Earth and are called **meteorites**. A single meteoroid burning up, creates a short quick streak of light and is called a **shooting star**. A group of them doing so, all at the same time, are called a **meteor shower**, and are named with suffixes -*ids* attached to celestial identifiers, which often are the constellations they occur in e.g. the **Perseids** in constellation Perseus.

Nebulae (or Nebulas) are interstellar clouds of material (gases and dust). Quite often they are 'nurseries' where new stars and their planets are born and 'raised', but they also can be the remains of dead stars, the smaller ones of which look like planets and are thus called **planetary nebulas**.

Novae (or Novas) are binary star systems whose interaction flares up, giving a brilliant appearance of a 'new star' but which eventually fades back to its previous state. Novas are rarely visible to the naked eye. Some novas result in super explosions called **supernovas**, resulting in the stars' destruction. Supernovas are rarely visible to the naked eye. But, supernovas are not just novas exploding to death. They are also phenomena associated with the deaths of super-giant stars, one of the most famous ones was observed a millennium ago, with its remnants now forming the **Crab** nebula. [Btw, when one views a nova or a supernova – through a telescope – one is witnessing an event that took place years and years ago! Or, from a different perspective, a star one views today may already have been destroyed – it's just that we won't know about it for years, generations, millennia, ...]

Pulsars are high-rotation neutron stars (made up of neutrons) pulsating electromagnetic radiation in a specific direction, detected on Earth if its radiation is beamed in our direction.

Quasars are galactic centres of very high lumosity, seen in telescopes resembling a bright star.

Before we take a closer, detailed look at the night sky in Part 2, the coordinate systems that are in use in astronomy have to be understood, as described in the next chapter.

6. Sky & Earth Graphs

Part of high school geometry includes graphs, consisting of a pair of axes, called x-axis (in the horizontal direction) and y-axis (in the vertical direction), and an origin point, where these axes intersect, which typically is assigned x and y values of zero and zero. Then, any point on the graph can be specified by (x, y) values e.g. (3, 4) specifies a point 3 units from the y-axis in the horizontal direction and 4 units from the x-axis in the vertical direction. Such x and y coordinate systems are called **2D Cartesian Coordinates** (2D for two-dimensional) as shown in the figure below.

2D Cartesian Coordinates

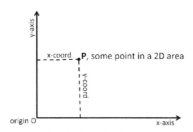

P's location is expressed as (x, y) e.g. (3, 4) which are in linear units such as metres

3 2D Cartesian Coordinates

The above (x, y) units are linear. In an alternate coordinate system called **2D Polar Coordinates**, a mixture of linear and angular values are used, as seen in the figure below.

2D Polar Coordinates

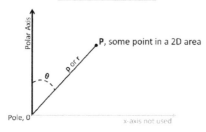

p (rho) or r is in linear units such as metres (called **Radial** distance)
θ (theta) is in angular units such as degrees (called **Polar** angle)
P's location is expressed as (p, ⊖) e.g. (5, 45°)

4 2D Polar Coordinates

The above systems cater to 2D areas. A space, on the other hand, is **3D** which is described in a three-coordinate system, as shown in the figure below, the third, additional dimension being specified by a z value, as in (x, y, z) e.g. (3, 5, 7) where the 7 indicates that the point is 7 units above the x/y area.

3D Cartesian Coordinates

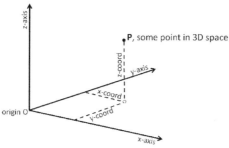

The three axes x, y and z are at right-angles to each other at the origin O.

P's location is expressed as (x, y, z) e.g. (3, 4, 7) which are in linear units such as metres

5 3D Cartesian Coordinates

Again, (x, y, z) units are linear. In an alternate coordinate system called **3D Spherical Coordinates**, a mixture of linear and angular values are used, as seen in the figure below.

3D Spherical Coordinates

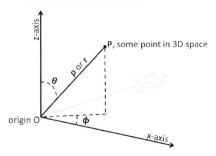

ρ (rho) or **r** is in linear units such as metres (called **Radial** distance)
θ (theta) is in angular units such as degrees (called **Polar** angle)
φ (phi) is in angular units such as degrees (called **Azimuth**)
P's location is expressed as (ρ, Θ , φ) e.g. (5, 45°, 30°)

6 3D Spherical Coordinates

Earth's Spherical Coordinate System

For our purposes herein, we use only the 3D Spherical Coordinates, but without the **radial distance**. [Instead of **radial distance**, which would specify the distance of point P from the centre of the earth, the height above mean sea level is used as the third value.] For example, Cornwall, Ontario, can be specified (using whole units in this example) as shown in the figure below.

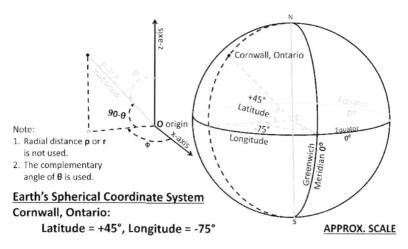

Note:
1. Radial distance **p** or **r** is not used.
2. The complementary angle of **θ** is used.

Earth's Spherical Coordinate System
Cornwall, Ontario:
    Latitude = +45°, Longitude = -75°

APPROX. SCALE

7 Earth's Spherical Coordinates

By convention, Latitudes are positive in the northern hemisphere and negative in the southern, and Longitudes are positive East of the Greenwich meridian and negative to the West.

Celestial Equatorial Coordinate System

The above system is used to locate points on Earth. For locating points in the sky, the above model is expanded almost infinitely to form a vast sphere called the ***Celestial Sphere (CS)***, on which all stars are imagined to be located. This is shown in the figure below, using star Capella as an example. This view is from the outside of the ***CS***, as compared to sky views further on herein which view the ***CS*** from the inside as that is how an observer on Earth would view it.

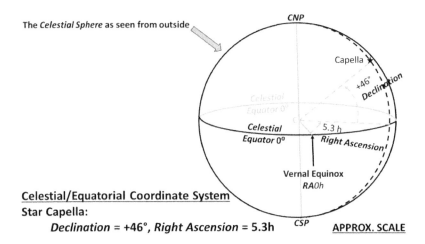

The *Celestial Sphere* as seen from outside

Celestial/Equatorial Coordinate System
Star Capella:
   Declination = +46°, Right Ascension = 5.3h   APPROX. SCALE

8 Celestial Coordinates

The above system is also known as the **Equatorial Coordinate System**, on account of it being based on Earth's Equator and its parallels (Latitudes) projected on to the **CS**. [Note: Earth's Longitudes are NOT projected since they would not be stationary lines on the **CS** (unlike Latitudes which are) because the Earth rotates. Instead, a fixed point on the *Celestial Equator* (*CE*) is used as the 0h, being the *Vernal Equinox*, also known as the First Point of Aries. This is the celestial equivalent of Earth's zero Longitude at Greenwich, England, and like Longitudes, the 'longitudes' of the CE are measured in the easterly direction, these 'longitudes' being called *Right Ascensions (RAs)*.]

Local Horizon Coordinate System

Finally, we need to be able to state the position of a sky object, for a specific date-time and location, using *local* references at the ground we are on. Such a system is called a **Local Horizon Coordinate System**, as shown in the figure below, again using star Capella as the example.

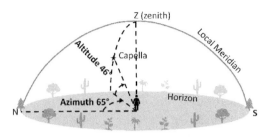

At Cornwall, Ontario, at 4 am Aug 21, 2018

Local Horizon Coordinate System
Star Capella:
    Azimuth = 65°, Altitude = 46°
                                            <u>APPROX. SCALE</u>

9 Local Horizon Coordinates

Given the celestial coordinates of a star and the date/time/location, we can convert them to Local Horizon Coordinates and vice versa, as seen in the examples in **Appendix B: Converting Celestial Coordinates to Local Horizon Coordinates** and **Appendix C: Converting Local Horizon Coordinates to Celestial Coordinates.** The beauty of the local system is that it does not involve any grid lines (e.g. x- and y-axes). It makes use of the local horizon along which to measure or lay out an arc, and at the end of which to measure or lay out an arc up the sky to the object of interest. The above figure shows this simplicity: from the North point on the horizon, a clockwise arc along the horizon is laid out for 65° and from that point another arc skyward is laid out for 46° and voila – star Capella. And for the case of a star you have observed but don't know what it is, you measure (not lay out) the degrees in reverse order: from the object in the sky to the nearest point on the horizon, thence along the horizon an anti-clockwise arc to the North point.

Bringing it all together – graphically

Here we will consolidate what we have learnt so far, by building the **CS**, a step at a time.

Grid Lines

We start with just the **CS** with Earth in its centre, as in the following diagram, drawing on it our first reference line, the **CE**, with its value of 0°.

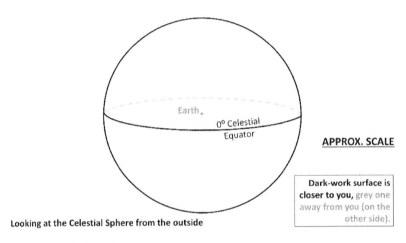

Earth

0° Celestial Equator

APPROX. SCALE

Dark-work surface is **closer to you,** grey one away from you (on the other side).

Looking at the Celestial Sphere from the outside

10 Earth inside the Celestial Equator

Now we add another great circle, labelling it as **RA12h** in the front, and *RA0h* at the back. We also label the sides as **RA6h** and **RA18h**. The top of this, and any other similar meridian, is always **CNP Decl90°**, while the bottom is **CSP Decl-90°**. Note that we can not label East nor West but we CAN say 12h is due East of 6h.

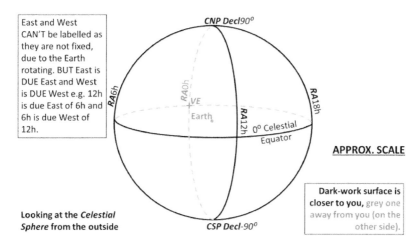

East and West CAN'T be labelled as they are not fixed, due to the Earth rotating. BUT East is DUE East and West is DUE West e.g. 12h is due East of 6h and 6h is due West of 12h.

CNP Decl90°

RA6h

RA0h

VE

Earth

RA12h

RA18h

0° Celestial Equator

APPROX. SCALE

Dark-work surface is closer to you, grey one away from you (on the other side).

Looking at the *Celestial Sphere* from the outside

CSP Decl-90°

11 Celestial Sphere with Main RAs

Before we plot our first star, we need more grid lines, as follows:

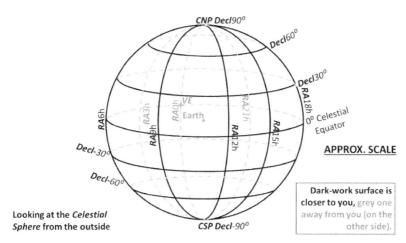

CNP Decl90°

Decl60°

Decl30°

RA18h

0° Celestial Equator

RA6h

RA3h

RA0h

VE

Earth

RA21h

RA9h

RA12h

RA15h

Decl-30°

Decl-60°

APPROX. SCALE

Dark-work surface is closer to you, grey one away from you (on the other side).

Looking at the *Celestial Sphere* from the outside

CSP Decl-90°

12 Celestial Sphere with more grid lines

Star Plotting

We are now ready to plot our first star! **Sirius the Dog Star**, the brightest in the sky. We place it by eye, as best as we can, labelling it "1" and showing its coordinates in a table: *RA6.75h*, being three-quarters of an hour ahead of our 6h edge, and *Decl-*

16.7°, just over half the distance between the *CE* and *Decl-30°*. [Star coordinates for these brightest ones are found in **Table of Coordinates: by Brightness** in **Appendix L: Information Tables**.]

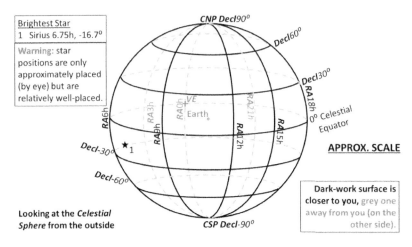

13 Celestial Sphere showing Sirius

Now we plot the next 20 brightest stars, these 21 in all constituting the complete set of stars of First Order **Magnitude**. These will be useful in finding our way in the night sky. The different sizes of the star symbols roughly reflect their **Magnitudes**. Notice there are four Zodiac constellation stars among the brightest, and by sheer coincidence, in one sequence viz. 14 Aldebaran in Taurus, 15 Spica in Virgo, 16 Antares in Scorpius and 17 Pollux in Gemini.

Having plotted all the brightest stars, we remove all the grid lines except for the main ones to reduce clutter:

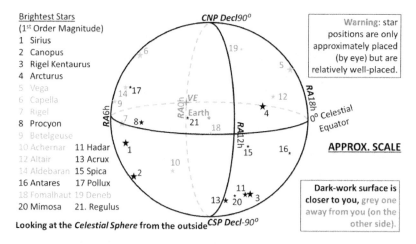

Brightest Stars
(1st Order Magnitude)
1 Sirius
2 Canopus
3 Rigel Kentaurus
4 Arcturus
5 Vega
6 Capella
7 Rigel
8 Procyon
9 Betelgeuse
10 Achernar   11 Hadar
12 Altair   13 Acrux
14 Aldebaran 15 Spica
16 Antares   17 Pollux
18 Fomalhaut 19 Deneb
20 Mimosa   21. Regulus

Looking at the *Celestial Sphere* from the outside

CNP Decl90°

Warning: star positions are only approximately placed (by eye) but are relatively well-placed.

APPROX. SCALE

Dark-work surface is closer to you, grey one away from you (on the other side).

CSP Decl-90°

14 Celestial Sphere showing 1st Order Magnitude Stars

We pause here to see how this view of the **CS** from the outside looks like from the inside as by an observer on Earth which is us!

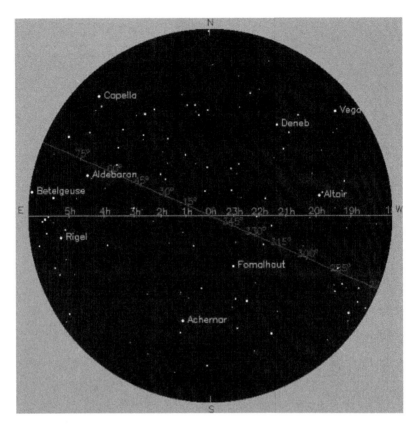

15 Lat0°, Long0°, Nov 1 2018, 21:15 UTC (from "Your Sky" by Fourmilab)

First, let's orient ourselves. Our position on Earth and the date-time of observation is given in the above caption. The horizontal line running through the middle of the screen shot is the **CE** with **RA**s marked out. [Note that the top of **Orion the Hunter**'s 3-star belt, between Betelgeuse and Rigel is just touching the **CE**.] The centre of the screen shot is our **Zenith**, which is **RA0h** as per the marking on the **CE**, and **Decl0°** as per the Latitude in the caption. In other words, from Earth (as in our **CS** diagram we have drawn thus far) we, meaning our **Zenith**, are looking towards the *RA0h* point on the **CE** on the far side of the diagram. We are looking at the *whole* back shell/dome of the **CS**, the edges of the shell/dome in view being the local horizon on Earth. Now we can match our plotted stars with the ones seen in the above screen shot. [One additional point on orientation: the screen shot is marked "S" at the bottom, denoting the South horizon. This means we are standing facing South and North is behind us. If you lift the above screen shot to your overhead, it then makes sense: "S" is pointing to the South horizon, "N" to the North, and to your left is the East horizon, to your right the West!] By the way, the line running on a slant through the screen shot marked in degrees is the *Ecliptic* which we have not yet marked in our above *CS* drawing.

The above screen shot, produced using Fourmilab's "Your Sky" software (see **Appendix F: Night Sky Tool**), can be re-centred to your Latitude / Longitude and date-time, and you are all set for observations. On the other hand, you may want to use our diagram, thus far, for a preliminary survey in order to choose the part of the sky that interests you. We continue with this objective now.

Zeniths = *Declination* Circle

Say, you're at Latitude 26° and Longitude -80°. [This is roughly the Atlantic beach area of South-East Florida, whereas the above zero-zero location of the screen shot would be on a ship in the Gulf of Guinea, off the coast of Gabon!] First, we draw the circle of *Decl26°*, being the projection of the Latitude onto the *CS*. This represents all the *Zeniths* in a 24-hour period of locations of all places on the Earth at this Latitude. So where is the *Zenith* at Longitude -80°? The answer requires the *Sidereal Time* of observation. The *Zenith* then is along *RA = Sidereal Time*. This can be calculated but it is readily available on the Internet or a smart phone app. For example, my smart phone app called "Sidereal Clock" tells me that at the moment:

The local *Sidereal Time* is 13:55:26 and the *Standard Time* (i.e. our usual clock time) is 7:56:00, a difference of 5:59:26. Say we want to make observations at 20:00:00 (8 pm). We add to it the above difference, getting the equivalent of 25:59:26 plus an adjustment* of about 00:02 giving us 26:01:26 = 2.0h rounded to nearest one decimal place. The **Zenith** then is along the **RA2h** meridian [*For very 24 hours of **Standard Time**, the **Sidereal Time** is ahead by about 3m 56s. For our naked eye astronomy, this adjustment can be ignored.]

Using the above simple calculation, we can calculate the **Zenith RA**s at sunset and sunrise, in between which is the time window of all **Zeniths** tonight. This is shown in our updated **CS** diagram below.

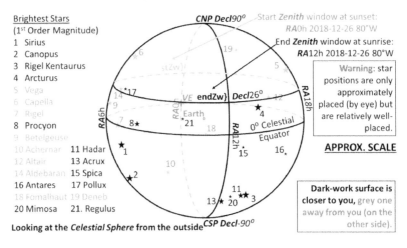

16 Celestial Sphere: a Time Window

The screen shot below is what the sky looks like at sunset, showing its reddish glow (in the eBook version) at the time. [Notice the **CE** is curved as compared to the previous screen shot where it was a straight horizontal line. This is because the **Zenith** of the former was along Latitude 0° whereas below it is along Latitude 26° and thus the **CE** is below it and comes out curved as per the rendering.]

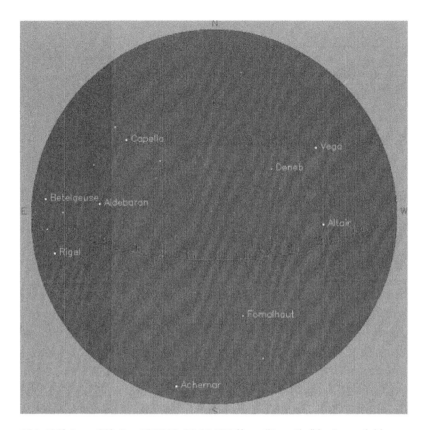

17 Lat26°, Long-80°, Dec 26 2018, 23:00 UTC (from "Your Sky" by Fourmilab)

Compass point labelled "N" is the northerly horizon in the above screen shot, whereas the **CNP** is marked "+", about an inch or so below it. How do the stars in it match up with our diagram at the sunset **Zenith**? To find out, we need to draw the horizon, as shown in the diagram below:

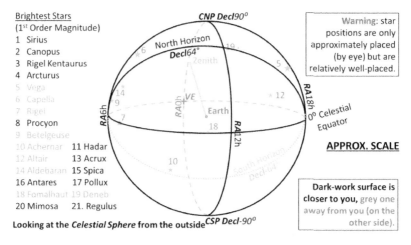

Brightest Stars
(1ˢᵗ Order Magnitude)
1 Sirius
2 Canopus
3 Rigel Kentaurus
4 Arcturus
5 Vega
6 Capella
7 Rigel
8 Procyon
9 Betelgeuse
10 Achernar   11 Hadar
12 Altair   13 Acrux
14 Aldebaran   15 Spica
16 Antares   17 Pollux
18 Fomalhaut 19 Deneb
20 Mimosa   21. Regulus

Looking at the *Celestial Sphere* from the outside

Warning: star positions are only approximately placed (by eye) but are relatively well-placed.

APPROX. SCALE

Dark-work surface is closer to you, grey one away from you (on the other side).

18 Horizon of the above screen shot & visible stars

Horizon

By definition, a local horizon is made up of all points 90° down the sky from the **Zenith**. Since in the above example we are at Latitude 26°N (which is equivalent to **Decl26°**), we see therefore the northern horizon is past the **CNP** on to **Decl64°** (90 – 26). Similarly, the southern horizon. The eastern and western horizons are the **CE**. Then, everything above the horizon is what we see in the sky at that time. This is seen in the above diagram where only the stars above the horizon are shown, and we can see that they match the previous screen shot.

The screen shot that follows shows the sky before sunrise:

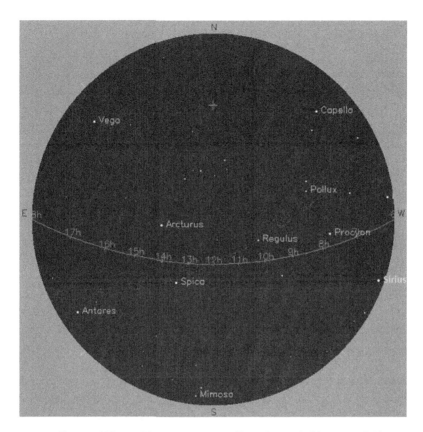

19 Lat26°, Long-80°, Dec 26 2018, 11:00 UTC (from "Your Sky" by Fourmilab)

Again, how do the stars in the above match up with our diagram at the sunrise **Zenith**? And again, we need to draw the horizon at the time, as shown in the diagram below:

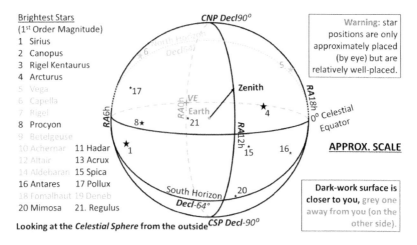

Brightest Stars
(1st Order Magnitude)
1 Sirius
2 Canopus
3 Rigel Kentaurus
4 Arcturus
5 Vega
6 Capella
7 Rigel
8 Procyon
9 Betelgeuse
10 Achernar   11 Hadar
12 Altair   13 Acrux
14 Aldebaran 15 Spica
16 Antares   17 Pollux
18 Fomalhaut 19 Deneb
20 Mimosa   21. Regulus

Looking at the *Celestial Sphere* from the outside

20 Horizon of the above screen shot & visible stars

Everything above the horizon is what we see in the sky at that time. This is seen in the above diagram where only the stars above the horizon are shown, and we can see that they match the previous screen shot.

The takeaway from the above is that where there is at least 12 hours of darkness, you will be able to view a complete rotation of the **CS** in one night, meaning you can observe ALL the stars visible at that Latitude e.g. you can see in the screen shot Polaris the North Star (marked by a "+"), however you can not the **CSP** and you can only see the top half of **Crux the Cross** (which is the southerly direction indicator, Mimosa being its second brightest). In other words, your northern horizon is 26° North of Polaris but in opposite sense the southern horizon is 26° *short* of the **CSP**. [Reminder: At any time our view of the sky is a half-shell (dome) of the **CS**, with the **Zenith** at the shell/dome's apex.]

Back to our diagram. Let's add a couple of things and wrap it up:

Ecliptic
(1) The **Ecliptic**. This requires plotting the Sun's apparent path across the **CS**. See **Table of Monthly Coordinates of the Sun** in

Appendix L: Information Tables which lists 2018 sun coordinates month by month. The direction of this apparent path is indicated by a solid black block-arrow in our diagram below.

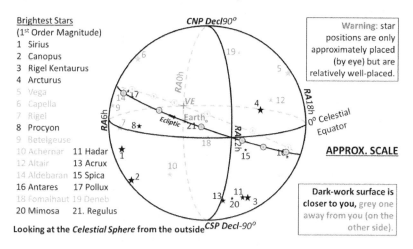

Brightest Stars
(1st Order Magnitude)
1 Sirius
2 Canopus
3 Rigel Kentaurus
4 Arcturus
5 Vega
6 Capella
7 Rigel
8 Procyon
9 Betelgeuse
10 Achernar   11 Hadar
12 Altair   13 Acrux
14 Aldebaran 15 Spica
16 Antares   17 Pollux
18 Fomalhaut 19 Deneb
20 Mimosa   21. Regulus

Looking at the *Celestial Sphere* from the outside

Warning: star positions are only approximately placed (by eye) but are relatively well-placed.

APPROX. SCALE

Dark-work surface is **closer to you,** grey one away from you (on the other side).

21 Celestial Sphere showing the Ecliptic

Milky Way Plane

(2) The **Milky Way** plane. This requires the celestial coordinates of its four cardinal points viz 0° (the galactic centre), 90°, 180° (the galactic anti-centre) and 270°, which are listed as "Gal. 0, 0" through "Gal. 0, 270" in **Table of Coordinates: by Common Name** in **Appendix L: Information Tables**. The plane that has been drawn in our *CS* represents the grey disc seen in figure **2 Galactic Coordinate System**. Not surprisingly, a majority of the 21 brightest stars are in or close to this band.

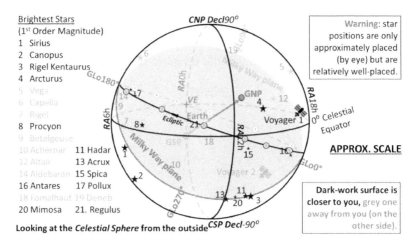

Brightest Stars
(1ˢᵗ Order Magnitude)
1 Sirius
2 Canopus
3 Rigel Kentaurus
4 Arcturus
5 Vega
6 Capella
7 Rigel
8 Procyon
9 Betelgeuse
10 Achernar   11 Hadar
12 Altair   13 Acrux
14 Aldebaran   15 Spica
16 Antares   17 Pollux
18 Fomalhaut   19 Deneb
20 Mimosa   21. Regulus

Looking at the *Celestial Sphere* from the outside

Warning: star positions are only approximately placed (by eye) but are relatively well-placed.

APPROX. SCALE

Dark-work surface is **closer to you,** grey one away from you (on the other side).

22 Celestial Sphere showing the Milky Way band

The various lines drawn in our above diagram will serve as orientation for the detailed tour in **7 A Tour of our Night Sky, Part 2**. [Notice that we sneaked in a couple of goodies in the above figure: NASA's space probes Voyager 1 and Voyager 2. Voyager 1 is currently about 145 AUs (1 AU = Sun-Earth distance) away from us and Voyager 2 about 120 AUs, after travelling for over 40 years (escaping Sun's gravitational pull but still retaining its orbit around the Sun, albeit now in a hyperbolic form, as it spirals out of our world in this fashion into the beyond). Being small and now far away, they are not visible but their signals are (barely) detectable by VLA (very large array) of radio telescopes. In about 40,000 years, both Voyagers are expected to pass by – within 1 to 2 light years – small stars, Voyager 1 passing by Gliese 445 in constellation Camelopardalis (presently in this constellation, but by then it will be located differently), Voyager 2 by Gliese 905 in constellation Andromeda (presently).]

Exercise: Reading *RAs* & *Decls*

A very useful exercise at this point, is to determine the **RA** and **Decl** of a star by eyeballing off the **CE** and the four **RA** lines in the above diagram, then checking this against the star's entry in **Table**

Ecliptic Coordinate System

Planets trace a nearly common path in the sky as they orbit the Sun. Earth's path is called the *Ecliptic* (strictly speaking, it is the Sun's apparent path in the sky, but which is actually caused by Earth's revolution around the Sun). It is convenient to have a coordinate system based on it for specifying locations of planets (and other solar bodies). It is structured similar to our **Celestial Coordinate System** (a.k.a. **Equatorial Coordinate System**) in that both have Latitudes / *Declinations* and Longitudes / *Right Ascensions*, the difference being the **Ecliptic** system is based on the **Ecliptic** plane whereas the **Equatorial** system is based on the **Equator** plane. But both are Earth-centred and both measure Longitude in the easterly direction from the *VE*. For example, the **Ecliptic** Longitudes today (December 27, 2018) of the Sun and planets which were in its apparent neighbourhood were as follows:

229° Venus
250° Jupiter
257° Mercury
275° Sun
280° Saturn

This tells us the order in which the planets vis-à-vis the Sun lined up in the sky: three planets to the right of the Sun, with Venus the furthest from it, Mercury the closest, and one planet, Saturn, to its left. Thus, just before sunrise, the three planets to its right are visible viz. Mercury, Jupiter and Venus, but Saturn is not since it comes after the Sun (Saturn will become visible at sunset where it will still be behind the Sun but in a then dark sky). After sunrise, only Venus is visible because being nearest to us it appears to be the largest and hence reflects that much more of Sun's light. [But once the Sun is high up enough and glaring, even Venus is not visible.]

Tip: All three coordinate systems viz. celestial, galactic and ecliptic are earth-centric, meaning their coordinate values are as observed from Earth.

Tip: Each of these coordinate systems provides a locational perspective which is different from each other. Also, these are perspectives of an environment which is in motion (apparent or real). Moreover, one's Latitude has an impact on these perspectives. As a result, it can be confusing in trying to orient and reconcile the various different perspectives (with the objective of staying on top of the big picture), especially in the early going. I can say this based on my own experience. My tip on this is, initially, not to worry about the big picture (I call it absolute orientation) at your observation location, because it requires much knowledge, which this book does provide but which has to be applied to field observations to fully grasp it. Instead, just use the information that is provided and explained in this book to the field work. Then, in time, the big 3D picture will emerge. In other words, we don't want to get ahead of ourselves, letting time and experience eventually lead us to the stage when we can look up at the sky, imagine the *CE*, find say **Orion the Hunter** (if he is in the sky at the time, otherwise some other *CE* constellations like **Zodiac** constellation **Virgo the Virgin**, or **Aquila the Eagle**, etc) and look for the hunter's dogs as they follow him, **Canis Major the Big Dog** and **Canis Minor the Small Dog**, in pursuit of the bounding **Lepus the Hare**, and then jumping right into the **Winter Hexagon**, a region of bright stars, knowing that we are looking through the **Milky Way** band (visible or not, but we know it's there) at our galaxy **Anti-centre**, and that the **Ecliptic** also cuts through the hexagon, ... This stage will come, I can assure you.

6 Sky & Earth Graphs

7. A Tour of our Night Sky, Part 2

Constellation Lyra the Harp: We look at this first, as it has star **Vega**, which is not only one of the brightest stars in the northern hemisphere (5th brightest overall) but is also one of the corners of the **Summer Triangle** of stars **Vega-Altair-Deneb**, in order of brightness (5th, 12th and 19th brightest in the sky). This is somewhat an isosceles triangle with Vega-Deneb the base – 2 ½ fists wide – and Altair at the peak below, the sides being just over 3 fists long. Although these stars belong to different constellations, viz. **Lyra the Lyre, Aquila the Eagle** and **Cygnus the Swan** respectively, they are great candidates for a would-be constellation called Summer Triangle, logically justifiable. But such non-constellation star patterns are known by another term, **asterisms**, to distinguish them from the conventional constellations.

There are two approaches to determining the current location of a star in a night sky for a given date-time and Latitude-Longitude of observation:

(1) A **coordinate** approach which gives us the star's *Altitude* (arc height above the horizon) and its *Azimuth* (arc distance along the horizon from its North point), used to locating it by the approximate method of using fist-n-thumb, or even by eye-balling.

(2) A **graphic** approach, using a rendering tool that displays the night sky, which then is matched to the actual sky to identify stars and constellations.

The coordinate approach is focussed on locating a specific star, whereas the graphic approach displays a whole bunch of stars that satisfy a given set of inputs. The first approach is formulaic, that is a star's location on the *Celestial Sphere* (this was dealt with in **Bringing it all together – graphically** in **6 Sky & Earth Graphs**) is converted to its relationship to the local horizon and the observer's *Zenith* which allows you to locate it in the local sky. The second approach being graphic, it therefore is user-friendly but requires trial-n-error input to reduce clutter and even then it requires experience in relating it to the actual sky in which the stars are NOT as bright as in the graphic.

Both approaches are presented here below:

(1) Co-ordinate Approach

Constellation Lyra the Lyre

Vega's celestial coordinates are: [obtained on the Internet by searching "star vega coordinates"; for your convenience, **Table of Coordinates: by Common Name** in **Appendix L: Information Tables** lists them]

Right Ascension (*RA*, equivalent to a celestial Longitude) **18h 36m 56s ≈ 18.6h**

Declination (*Decl*, equivalent to a celestial Longitude) **38° 47′ 01″ ≈ 39°**

Say, we are observing a night sky at the following earth coordinates (Cornwall, Ontario): [You can obtain your local Latitude-Longitude from say Google Maps by clicking on your location: a pop-up will display the Latitude and Longitude; they are also displayed in the web address near the top e.g. "Cornwall,+ON,+Canada/@45.0451978,-74.8268962"]

Latitude **+45°** (i.e. half way between the Equator and the North Pole)

Longitude **-75°** (or **-5h**, 1 hour of Longitude = 15° of Longitude)

Assume we know where the *CNP* (*Celestial North Pole*) is – it's easy to find, as we shall see later. Because the Earth rotates, West to East, around its North-South axis (whose projection on to the *Celestial Sphere* of stars yields the *CNP* and *CSP*), so do the stars appear to rotate in the opposite direction, East to West, around the same axis, rising in the East, setting in the West, just like the Sun. When a star, by this apparent rotation, crosses our *Local Meridian* (*LM*, more below), it is said to be in **transit**, and this is the highest it will be above the horizon, halfway between its star-rise and its star-set. This transit time is easy to calculate given a star's coordinates, such as for Vega above, as we see below.

Our **LM** (which is the Longitude of our observation point projected on to the **Celestial Sphere**) is constantly moving because of Earth's rotation. [Note: the Longitude itself does not change, as it is fixed with respect to the zero at Greenwich, and thus its location relative to the Earth is stationary.] At any point in time, the **LM** is its **RA** at that exact point in time, by definition. It is also equal to the Local **Sidereal Time** (LSidT) since both **RA** and LSidT use the same zero celestial meridian and both are expressed in **Sidereal Time**.

The LSidT at the above Longitude of observation at 9 pm on August 30, 2018 was: **18h 37m** (obtained by searching "sidereal time calculator" on the Internet and filling in a form – see example in **Appendix A: Obtaining Local Sidereal Time**) which is seen to be equal to Vega's **RA**, thus the **LM** and Vega's **RA** being coincident, Vega was **transiting** at that time. [The reverse example would be that, knowing the **RA** of a star, we can use it as the LSidT of its transit and at that time of the LSidT, observe it in our **LM**, (that is, as long as the LSidT is in night time making the star observable, otherwise a different day has to be considered).]

For our date-time of observation, Vega was at the following **Altitude** above the horizon, while transiting the **LM**:

90° – absolute (observation Latitude – Vega's **Declination**)

= 90 – absolute (45 – 39) = **84°**

which makes it 6° to the South of **Zenith**. Thus, the brightest star of the Summer Triangle is identified for this particular locale for a specific date and time.

But, Vega's off-transit locations, at different times, are not so easy to calculate. However, it is easy if a calculator on the Internet is used. See **Appendix B: Converting Celestial Coordinates to Local Horizon Coordinates**. Say, you want to locate **Vega** in the sky tonight (say, November 6, 2018, at 9 pm) at the same location as above (Cornwall, Ontario). Using the above appendix, we get the following **Local Horizon Coordinates** (LHC) to locate Vega in the sky:

Azimuth: 295° [approximately WNW on the horizon]

Altitude: 31° [degrees above the horizon from above *Azimuth*]

[These terms were covered in **6 Sky & Earth Graph**, and are also explained in the **Glossary**.] The above degrees can be eye-balled in the sky or laid out with fist-n-thumb.

The rest of the Lyre's stars are in the neighbourhood of Vega, as in the figure below. [The box around each stick figure is meant to clearly separate one figure from another and to serve as a labeller for the constellation which the figure represents. Specifically, the boxes are NOT the IAU constellation boundaries which are elaborate right-angled polygons. See figure **90 Chart of Orion the Hunter (by the IAU)**, for example.]

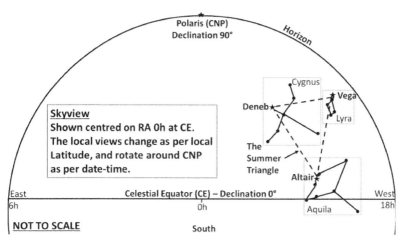

23 The Summer Triangle

Diagram-to-CS Orientation

This is straight-forward: just match the grid lines common to both the above chart and our previously developed **CS** model. The above chart is looking at the upper half of the rear of the **CS**, as shown by the grey hatchings in the figure below:

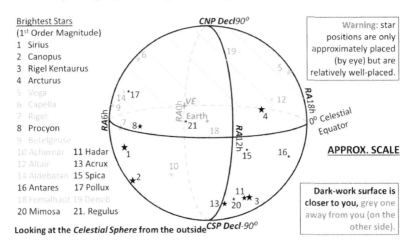

24 Diagram-to-CS Orientation

As is seen in the above model, the Summer Triangle stars #5, #12 and #19 approximately* match their representation in the previous chart. [*There is a difference between these two diagrams. The **CS**'s perspective in the above diagram is **3D**, whereas in the previous chart it is **2D**. Mathematical transformations are used to draw a 3D surface on a 2D medium such as paper, a very good example being an Earth map (2D) representing the spherical Earth (3D). Inevitably, this process causes some distortion e.g. a rectangular farm may not have exact right angles on its map, or its acreage on ground may not match that which is calculated from its paper map, etc. Mathematical formulae developed over the past centuries try to minimize one or more such distortions.]

Diagram-to-Sky Orientation

How do we relate a figure like the above to the night sky view? In two steps.

Step 1. First, we make note of the cardinal points in the figure e.g. in the above, the centre-top is showing star Polaris, which marks the **CNP**, the bottom-centre the South direction, East is towards left, West towards right (here, East and West are the opposite of what's found on a road map). Next, imagine you are facing the South horizon, matching its bottom-centre label. [South is either you know where it is locally, or you determine it as shown further below.] Raise the diagram above you, it facing you squarely down. Thus, the top of the diagram will be facing North (behind you), which matches the top-centre label. With you facing South, that too matches the bottom of the diagram. [A useful analogy for this is of reading a book in bed, with your feet-to-head line pointing in a northerly direction, and instead of resting the book on your chest you raise it overhead to read it.]

Step 2. This step is tricky, for it has to allow for the rotation of the sky around the **CNP**. The stars in the above diagram are plotted using their celestial coordinates, with the **CE** (with **RA**s marked) shown as being parallel to the bottom edge, which is typically how such diagrams are drawn. However, at any given time of the night, these stars could be anywhere around the **CNP**. The Summer Triangle in the above chart is seen near the West edge. But on any given night, the triangle may be located in the sky say near its East edge, or in the middle, or not there at all, because the stars are in continual rotation. For example, the triangle in some night sky along the **CE** might look like in the following diagram:

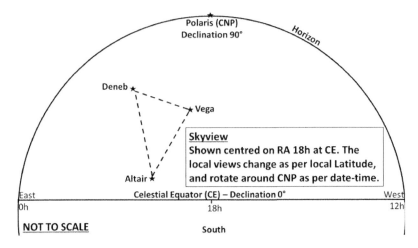

25 Summer Triangle – re-charted on RA18h

The above chart is the same as the previous except this one is centred on **RA18h**: we can see that the triangle is about 1h east of **RA18h**, same as in the previous chart (because although the stars rotate, their coordinates remain the same). But charts are standard, typically centred on **RA0h** or **RA12h**. **Step 2** of the orientation deals with orienting a standard chart to the current night sky.

If you have already located the Summer Triangle, this orientation is easy. Just rotate the chart around its **CNP** (if you actually have held it up as described in Step 1, or imagining it so) until the chart's triangle-centre is super-imposed on the sky's e.g. for the above example, you would rotate a standard chart to the East.

Conventionally, diagrams show a hemisphere (or half of it like the above), using **CE** scale lines centred on **RA0h** and for the other hemisphere, **RA12h**. But out in the field during observation, the imaginary **CE** line in the sky can be centred on any **RA**, like say the above diagram which is centred on **RA18h**. But what if you have not located the Summer Triangle yet. Is it possible to orient the chart without it? Yes, it is. [Which is equivalent to asking, where in the sky do you look for the triangle using a standard **RAOh**-centred diagram?] For this, you have to know what your current LSidT is, which is 18:37 (≈ 18.6h) as previously seen above. Thus, it is 5.4h *before* **RAOh** (0 – 18.6 + 24). [*'before'* as in the same way we tell relative times of our clocks, except here it is **Sidereal Time**: Thus, I look up the time in my sidereal clock (a smart phone app) and it says 18:37 and I see that my chart is centred on 0h. Ergo, I have 5:23 to go until 0h/24h. That's when my **LM** will be 0h, same as the chart!] From the overhead (i.e. the **Zenith**) to the East horizon is 90°, which is equal to 6h, meaning the 0h meridian is just above the horizon (i.e. in the process of rising off the eastern horizon). [Reminder: **RA**s are measured in an *easterly* direction, meaning their values increase in that direction, while the sky rotates in a *westerly* direction, meaning the **LM's RA** increases as the sky rotates.] Thus, the diagram that you had squarely above you (or imagined it as such) needs to be rotated, around the **CNP**, towards the eastern horizon until its 0h is just above the horizon. Now, it's oriented and you would look for the triangle accordingly.

Admittedly, Step 2 is not a piece of cake, but in time, with experience, it will become routine. The takeaway here is that a standard chart is like a snapshot of the sky, drawn with the **CE** line horizontal, and at an *arbitrary particular moment of time*, typically with the **RAOh** or **RA12h** line in the centre: thereafter, the actual sky continues to rotate, with its **RAOh** or **RA12h** line sweeping away to the western horizon.

Here is an additional perspective that helps us understand the orientation. If we extend the previous diagram to cover the whole of the rear shell of our **CS** model, we get:

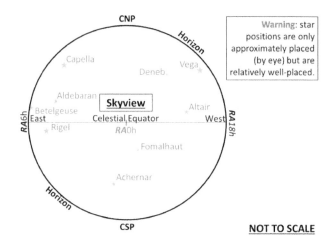

26 Full Coverage of Rear Shell of the CS

The above diagram covers the whole of the rear shell of our **CS** model. It is to be compared with figure **23 The Summer Triangle** whose coverage is only of the northern part of that rear shell, with the triangle's three stars labelled which the above diagram also has, but along with them, it shows all the 1st Order **Magnitude** stars of the rear shell. Assume you are Latitude 0° (or very close to it), matching the **CE** of the diagram. [At greater Latitudes, the local North horizon drops down from the North horizon of the diagram by an angular distance equal to the Latitude. In order words, star Polaris, representing the **CNP** will not be at the horizon, as in the diagram, but above it by a distance equal to the Latitude. This tilt of the horizon vis-à-vis the diagram's horizon is described with the aid of drawings in **Horizon** under **Bringing it all together – graphically** in **6 Sky & Earth Graphs.**] Now, if you were facing the South horizon, and you held the above diagram in front of you and began observing the sky above this horizon, you would first see star Achernar, then Fomalhaut, then Rigel / Betelgeuse / Aldebaran just above the East horizon, etc just as shown in the diagram! If instead, you were facing the North horizon, you would need to completely rotate the above diagram so that it's **CNP** label was at the bottom and the **CSP** at the top. Then, the order in which you would observe the stars above the North horizon would be Capella just above the northeast horizon, then Vega above the northwest horizon, etc. The takeaway from all this is that, a diagram like the above is properly oriented if the horizon you are facing matches the horizon at the bottom of the diagram.

Having said all that, you have an easier alternative, the graphical approach discussed further below in this chapter in **(2) Graphic Approach**, which allows you to produce a customized image of the local sky, with its field of view matching the horizon you are facing! No orientation required, just face the horizon point of your choice and use a rendering produced for what you are seeing! See figure **80 Customized Observation Screenshot (by Stellarium).**]

While we are on the subject of orienting diagrams, let's consider 3D diagrams / models (think of a globe in a geography class) e.g. of Earth, the Solar System and most importantly the **CS**. These are viewed from the *outside* of the model (e.g. a classroom globe), as compared to sky drawings like the above which are views from *inside* the **CS** looking UP to its inner surface, being the sky. These perspectives impact certain things, a significant one being the stellar rotation around the celestial poles. From the inside, the stars rotate around the **CNP** in an anti-clockwise direction i.e. in a *westerly* direction, and around the **CSP** in a clockwise direction (but still in a *westerly* direction). But when the **CS** is viewed from above the **CNP**, the stars rotate around the **CNP** in a clockwise direction (opposite to the inside view, but still in a *westerly* direction). Similarly, when the **CS** is viewed from below the **CSP**, the stars rotate around the **CSP** in an anti-clockwise direction (opposite to the inside view, but still in a *westerly* direction). [By the way, **RAs** and **Sidereal Time** are both measured in an *easterly* direction, just as Earth Longitudes are.] An additional comparison is between our sky diagrams and road maps. A road map is easily oriented with the ground that we travel/walk on, because when a paper map is unfolded on a desk/table it already is parallel to the ground that it is representing. A screen map too is relatively easy to orient. In both the paper and screen maps, North is typically at the top and thus East is to the right, West to the left. But when it comes to sky diagram (say in a book like this), it is not intuitively oriented (because it is read in a book on the table while the sky it is representing is overhead), and – most importantly – when North is at the top, East is to the LEFT, West to the RIGHT – opposite of a road map! See figure **80 Customized Observation Screenshot (by Stellarium)**.]

Sight Adjustment

TIP: Looking at the night sky, especially in light-polluted areas, requires an adjustment from the lit ground-level to the relatively darker sky. This is done simply by staring up for at least half a minute before attempting to observe. The eyes need this time to adjust to the darker sky. Then, make one of your observations while staring up all the time (so that ground lights do not un-adjust your earlier eye adjustments). Ideally, the ground-level lights need to be blocked out. Observatories do this by building a silo with a dome roof that opens a slit for a telescope below to observe through (being in addition to building it on a remote hill/mountain; Antarctica would rank up at the top of the remote list). But that is not an option for us wandering amateurs. You could try improvising e.g. using a manila board rolled up as a cylinder and some cloth to fashion something similar to the first still-cameras of the past where the photographer covered his head with a black head-covering cloth attached at the peering end of the camera! Or, a cardboard box, open at both ends, with shoulder cut-outs. Or, any similar improvisation. But how does then one make field notes? On my part, I just find a dark spot and let my eyes get used to the sky, cupping my eyes with my palms as needed.

Constellation Cygnus the Swan

We look at this next, as it is a neighbour of Vega and contains **Deneb**, a bright star, one of the corners of the **Summer Triangle**, and having located Vega, we just need to look for another bright star nearby. But it is one of *two* bright stars in the Summer Triangle, so we need to identify Deneb correctly. It is the least bright in the triangle, so that may help. Also helpful is that Deneb is closer – 2 ½ fists away – whereas Altair is 3 ½ fists away. But to be sure, we determine its Local Horizon Coordinates, same as was done above using an Internet calculator:

Azimuth: 289° [between West and WNW on the horizon]

Altitude: 55° [degrees above the horizon from above point]

The rest of the Swan's stars are in the neighbourhood of Deneb, as relatively shown in the previous figure **23 The Summer Triangle**. As the note in the figure alerts us, local views change, as we see above by the calculated **Local Horizon Coordinates**. Compare Vega, seen at the lowest altitude (distance to the nearest horizon point) in the triangle in figure **23 The Summer Triangle** (centered on *RA0h*), and figure **25 Summer Triangle – re-charted on RA18h**, where it's Deneb that is at the lowest altitude!

Interestingly, Deneb's Galactic Coordinates (GCs) are: [see **Appendix D: Celestial & Galactic Coordinates of Stars** for how to obtain them]

Galactic Longitude (Glong): 84°

Galactic Latitude (Glat): 2°

This means it is located in the Milky Way band (as per its Glat) and not too far from the 90° direction (as per its Glong). [More on the Milky Way band further on in this tour.]

Constellation Aquila the Eagle

We now complete the **Summer Triangle**, looking for **Altair**, the second brightest star in the triangle. Having already identified its two other stars, their nearest bright star would have to be it. Also, it is at the apex of the isosceles Summer Triangle. But to be sure, we determine its Local Horizon Coordinates, same as was done above using an Internet calculator:

Azimuth: 257° [between West and WSW on the horizon]

Altitude: 25° [degrees above the horizon from above point]

The rest of the Eagle's stars are in the neighbourhood of Altair, as relatively shown in the previous figure **23 The Summer Triangle.** Of these, the eagle's lower (southerly) wing tip just crosses the **Celestial Equator** (CE), while its tail hangs even further South.

In time, the triangle will be readily recognized by itself, with no need to use *Azimuths* and *Altitudes*. Here is a photo hot-off-the-press I took of the triangle:

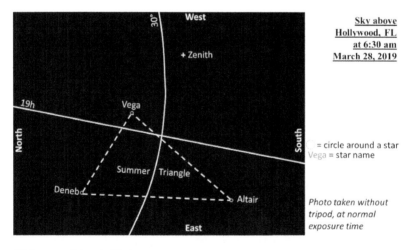

27 Summer Triangle Photo

For more on this and nearby constellations see **Observations at Collingwood, Ontario, Canada late June 21, 2015** in **11 Light Pollution Buster**, showing photo identifications, where it can be appreciated how bad light pollution is.

Constellation Orion the Hunter

This is one of the most recognizable constellations, with its iconic 'belt' of three stars close together at its centre, the top of which almost touches the *Celestial Equator* (*CE*). [Another important feature is that it has TWO First-Order Magnitude stars (Rigel and Betelgeuse), making it one of only three constellations having two such stars, the others being Centaurus (Rigel Kentaurus and Hadar) and Crux (Acrux and Mimosa).] Whereas the previous three stars Vega-Altair-Deneb formed a triangle, Orion's belt stars form a short straight distinct line. See figure below:

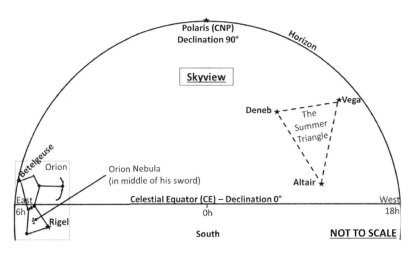

28 Orion

Once its belt is recognized, its other stars can be identified as per their locations in the above figure. It has two of the brightest stars, Rigel (7th), on the front knee of the hunter, and Betelgeuse (9th), on his rear shoulder, but which can be correctly identified by their Local Horizon Coordinates for the above observation point and date-time:

Rigel: Azimuth 105° [ESE], Altitude 3° [just above horizon at ESE]

Betelgeuse: Azimuth 97° [just South of East], Altitude -3° [just below horizon, so it can't be seen at this time, but should be visible in half an hour, after its star-rise]

Here is a photo I took of Orion above our townhouse, a decade and a half ago, in the twilight (3 belt-stars near the centre):

Constellation Orion

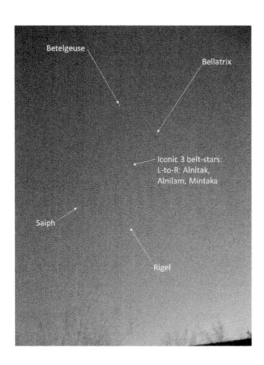

Sunday 7:30 pm, March 28, 2004, Toronto, ON

29 Orion photo

Orion not only fascinates with its most recognizable layout but has a bonus – the **Orion Nebula**. As pointed out in the above diagram, it sits in the middle of the hunter's sword. Nebulae are either nurseries (as in where children / plants are raised) of new stars or graveyards of dying stars. The Orion Nebula is of the former type. It is visible to the naked eye as a fuzzy, whitish patch, albeit only in a dark sky or very high up in the sky away from the lights of the horizon. Binoculars would help. A telescope would certainly show its colourful beauty.

Constellations Ursa Major & Minor, the Great and Little Bears

This pair is one of the most important navigational constellations, the North star Polaris being part of the latter. Each of these constellations has a well-known asterism, the Big Dipper (ladle) in the former, the Little Dipper in the latter. Although the North Star Polaris is in the latter, the bigger, brighter former is much easier to pick out in the sky. See diagram below. Look for the dipper (ladle) shape of the Ursa Major in the northerly direction, which is halfway along the horizon between the local sunrise (easterly) and sunset (westerly) points on the horizon. The big dipper's outer edge (the stars **Merak-Dubhe)** points to Polaris, about a distance slightly longer than **Merak-Alkaid**). Polaris should be at an altitude equal to the Latitude of the observation point (45° in our case). Why? Because if you stood at Earth's North pole (Latitude 90°) and looked straight up (i.e. altitude 90°) you would see Polaris, and if you were at any point on the Equator (Latitude 0°), Polaris would be barely visible at the horizon (altitude 0°) and ergo, in between, altitude = Latitude!

As seen in the figure below, Polaris is at the end of the little dipper/ladle handle asterism. An arc traced from your zenith "Z" to Polaris "P" and continued to the horizon gives you the local North point "N" on the horizon. If you continue this arc from Z the other way to the horizon, it gives you the local South point "S". This great arc N-P-Z-S is your *Local Meridian (LM)*, being the projection of the local Longitude on to the **CS**. As the Earth rotates so does the **LM** move on the **CS**. An arc distance of 90° from P along this **LM** in the southerly direction brings you to the *Celestial Equator (CE)*. The top of **Orion the Hunter**'s belt almost touches the **CE**, as shown in the diagram above.

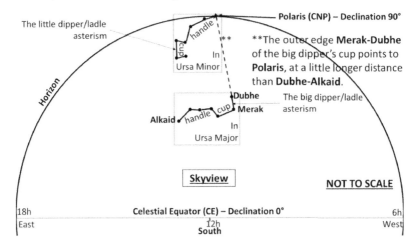

30 Finding the North Star Polaris

Constellations of the Milky Way Band

As we saw above, the Milky Way band cuts through constellation Cygnus. It turns out that the band arcs right through the middle of our observation area of the sky, as seen in the figure below, showing some more constellations, **Auriga the Charioteer, Cassiopeia the Queen of Ethiopia, Cepheus the King of Ethiopia** and Zodiac constellation **Taurus the Bull.**

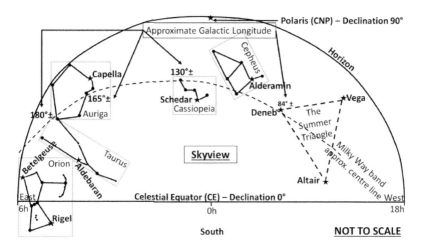

31 Milky Way Band Constellations

The galactic plane's anti-centre (180°) is near the tip of the Bull's northern horn. 90° is somewhere between Deneb and the bottom of Cepheus, closer to the former. The centre (0°) though is off this figure, as is 270°.

Planets & Zodiac Constellations

Now we come to bodies (planets) within our star system (our solar system). The closest or biggest planets are brighter than the brightest star in the sky. Venus, for example, is brighter than the brightest star **Sirius the Dog Star**. Although planets do not generate their own light (as stars do) they reflect our star the Sun's light off their relatively large surfaces (being very close to us, as compared to the distant stars). Although bright, the reflected light does not twinkle as starlight does (because stars are tiny pinpoints that are affected more by atmospheric attenuation than the larger, closer planets).

Just like Earth, the other planets also revolve around the sun but at different circuit times (one year for Earth) e.g. the closest planet to the sun, Mercury, takes 88 Earth days to complete its revolution, while the furthest one, Neptune, takes 165 Earth years. The reverse is the case for the planetary travel speeds around the sun: Mercury travels at 170,000 kph, while Neptune at 20,000 kph. The net result is that planets are not stationary in the sky with respect to stars and to each other. For example, Mars is visible at the observation point and date-time indicated at the bottom-right in the figure below; Uranus and Neptune are also within field-of-view but can't be seen by eye:

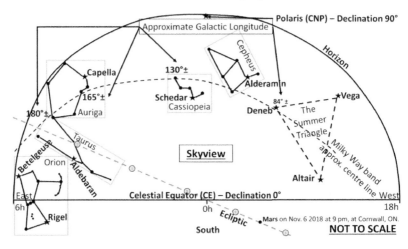

32 Mars

But a year after the date-time indicated it will no longer be in view while all the stars remain in the same locations in our field-of-view. To appreciate how planets become observable or not, look at the figure below, for the Mars observation date above:

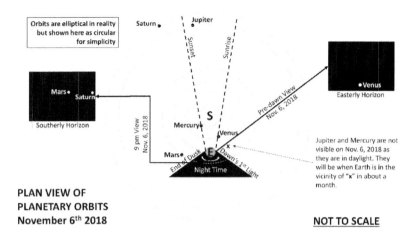

PLAN VIEW OF
PLANETARY ORBITS
November 6th 2018

33 Planets in View Nov. 6, 2018

Planets in daylight (Jupiter and Mercury in the above figure) are not visible on that date. Even Saturn is only clearly visible from sunset to just after 8 pm, after which it begins to set.

To observe planets other than Mars as seen above, we would have had to be observing exactly three months before the above date-time. Which brings up the question as to how do we know when we can observe planets? The answer is not straight-forward, but we can state something that is simple. All planets and the sun travel approximately on a single apparent path through the sky. This is because their orbits around the sun are close to coplanar, meaning their orbital planes are close to coincident. This travel path is called the Ecliptic, shown as a dashed line (an arc actually) passing through Mars in the previous figure **32**. But, it should be noted that this Ecliptic as seen against its celestial background is an apparent sinusoidal path up-above-and-down-below the *Celestial Equator*, with its upper peak just above and to the left of Orion, the lower peak being just to the left of Scorpius in the figure below, which also shows three observable planets, at the date-time indicated, plus some more constellations: **Bootes the Herdsman, Scorpius the Scorpion, Virgo the Virgin** and **Gemini**

the Twins – the last three being of special interest because they are Zodiac constellations; more on this later.

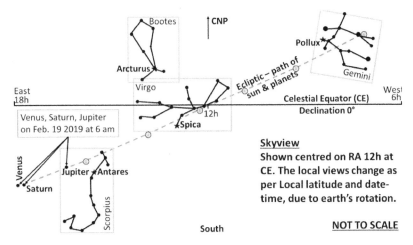

34 Venus, Saturn, Jupiter

Here is a photo I took covering the lower left corner of the above diagram: [Venus and Saturn are seen in a 1+° Conjunction (meeting)]

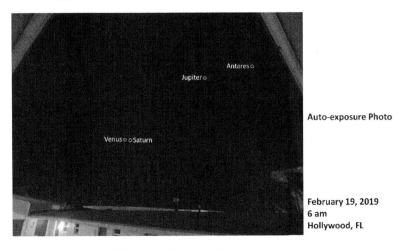

35 Photo of bottom-left corner of previous diagram

A couple of weeks later, I took another picture of the same night sky:

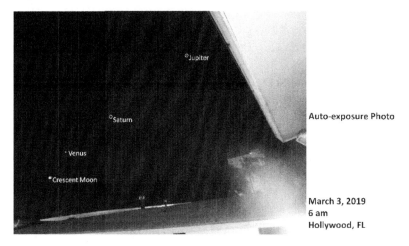

_{Jupiter}

Auto-exposure Photo

_{Saturn}

Venus

Crescent Moon

March 3, 2019
6 am
Hollywood, FL

36 Same sky shot as previous, couple of weeks later

The three previous planets are nicely arrayed here on the Ecliptic, seemingly in pursuit of a Crescent Moon. Venus does not require a circle around it as it can be readily seen. The Moon when present, even a crescent one, is always the brightest object in a night sky.

Again, to appreciate how these planets are visible, look at the figure below for the lay of the planets for the observation date-time:

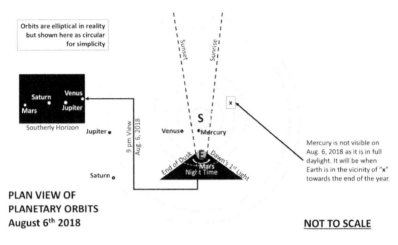

37 Planets in View Aug. 6, 2018

Here too one planet (Mercury) is not visible.

Back to the question of how to determine when we can observe planets. Although there are calculators on the Internet that give planetary coordinates for any given date and time, you still have to plot them against your night sky to see if they fall within view. If not, then you have to try another month, and so on. Very tedious. A better approach is to use a night sky rendering tool that graphically shows you the sky for a given date-time and Latitude-Longitude. This is still a trial-n-error approach as you have to try different date-times but its value is in not having to plot and re-plot planetary coordinates. This is tackled below under **(2)** Graphic Approach.

Tip: When we see a very bright 'star' it usually is a planet, and most likely it is one of the two brightest ones, Venus or Jupiter, in that order of brightness. As to which one it is, the easiest way is to use a rendering tool, such as in **Appendix F: Night Sky Tool**, and matching it by eye to the display (or more accurately, by measuring its Local Horizon Coordinates, converting it to *RA-Decl* and comparing this to what the rendering tool's grid says). If both are in view, the obviously brighter one of the two is Venus. On the other hand, Mercury, Mars and Saturn are in the same brightness category as the top ten bright stars, and so can be mistaken for stars. With observation experience, especially knowing where the Ecliptic is, you can rule out the possibility of a star being one of these planets.

One last point to tackle is the reverse to the above (which was preparatory to looking for stars and planets): what if you want to just dive right into observation, find something interesting up there such as the bright stars that you see? Then the quest is to identify them. You do this by measuring their *Azimuths* and *Altitudes*, then using a calculator on the Internet to determine their celestial coordinates, *RAs* and *Decls*, such as the example in **Appendix C: Converting Local Horizon Coordinates to Celestial Coordinates**. Now you can look up a table on the Internet of brightest stars sorted by *RA* and *Decl*. An example is the **Table of Coordinates: by RA-Decl** in **Appendix L: Information Tables**. A less tedious approach is a graphic one, given at the end of the section below:

(2) Graphic Approach

In this approach, a rendering tool is used to display the night sky for a given date-time and Latitude-Longitude of observation. A couple of examples of such a tool is given in **Appendix F: Night Sky Tool**. Using it, here below is an image for our observation location and date-time:

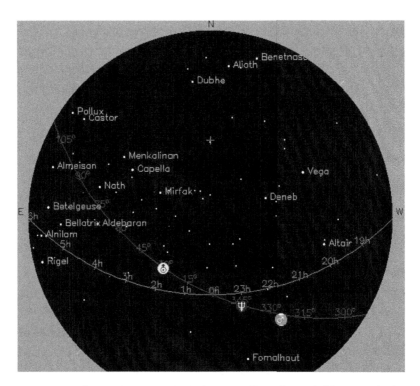

38 At Cornwall, Ontario, on Nov 6, 2018 at 9 pm (from "Your Sky" by Fourmilab)

To relate this to the previous figures, first locate the Summer Triangle stars Vega-Altair-Deneb. This anchors the right side of the previous figures. Next locate Betelgeuse and Rigel, the back shoulder and the front knee of Orion the Hunter, respectively. This anchors the left side of the previous figures. Now the rest of the features can be related, including the *CE*'s arc (marked in hours "h"), the Ecliptic's arc (marked in degrees °) and planets Mars (shown with its iconic male symbol), Neptune (with his Trident) and Uranus (with its Mars-like symbol turned 45° anti-clockwise, arrow pointing up) – although, to the naked eye, only Mars is visible. [The *CE* is shown as a straight line in the previous figures, but as an arc in the above image. This is as a result of a choice of a Latitude of projection, being the *CE* in the previous figures and the observation Latitude in the above image.] The input parameters for this graphic were set to show only the bright stars with star names labelled for the brightest only.

So, where are the other planets? We use the knowledge that the planets (as well as the sun) travel along an almost common path, the Ecliptic, shown above as an arc marked in degrees °. [But a quick aside here: It should be noted that they do not move like a Pied Piper (the sun) leading the rats (the planets), because they travel at different speeds. Thus, they appear to overtake one another – sometimes with one eclipsing the other in doing so.] ALL planets are out there in our sky, spread unevenly along the Ecliptic. Then, as the Earth rotates, it progressively sees all of the local sky within a complete rotation in 24 hours, divided into night and day (the latter makes stars and almost all planets invisible, Venus being the exception sometimes, when it is at its closest to Earth). For example, if we had observed two hours earlier (that is, 7 pm), we would have viewed two more bodies, Saturn (with its astrological symbol of a sickle) and dwarf planet Pluto (showing Percival Lowell's monogram), as seen in the rendering below.

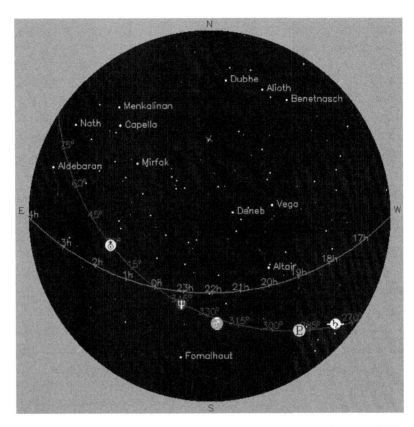

39 At Cornwall, Ontario, on Nov 6, 2018 at 7 pm (from "Your Sky" by Fourmilab)

But the rest of the night does not bring up any more planets into view, and which only come up in the daylight sky around mid-morning, thus not visible (except perhaps Venus, faintly, and then in the early, less bright, morning). This view is seen in the image below, where the day sky is rendered blue (in the eBook version) as compared to black in the night sky above. Note the location of the Sun (at 225°) and a crescent Moon (at 210°) on the Ecliptic.

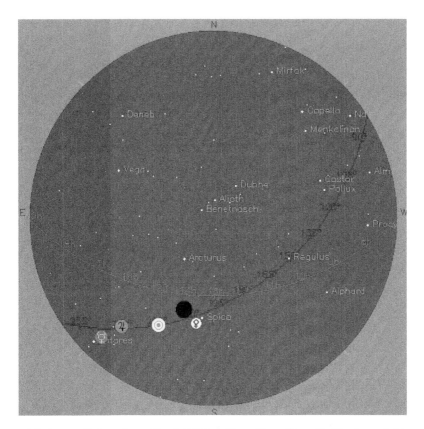

40 At Cornwall, Ontario, on Nov 7, 2018 at 10 am (from "Your Sky" by Fourmilab)

To view the planets which come up currently at day time, we would need to choose an observation date when they come up at night. Since the Earth goes around the Sun in one year, we should be able to view these planets in a timeframe of roughly half a year later or earlier. Thus, for the same day and time but month May, two of these planets are visible, viz. Jupiter (with its astrological symbol of Zeus' first letter, with a stroke through it) and Venus (with its female symbol), as seen in the image below.

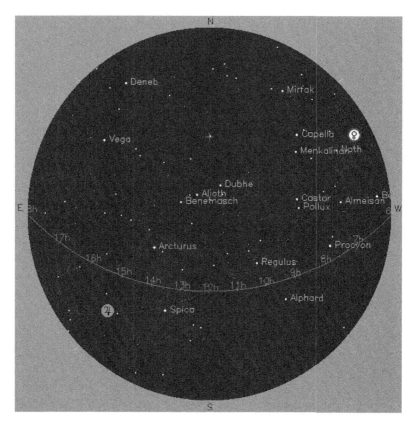

41 At Cornwall, Ontario, on May 6, 2018 at 9 pm (from "Your Sky" by Fourmilab)

But Mercury ain't there. Mercury travels at the highest speed of all planets, in a tight circle around the sun. Being close to the Sun, it is only visible just before sunrise (if it is ahead of it on the Ecliptic) or just after sunset (if it is behind it on the Ecliptic). Thus, the sky at sunset before our original date-time (see image below) shows Mercury (with its Venus-like symbol, wearing the winged cap of god Mercury) trailing the setting Sun, plus a small bonus – a total of six planets plus dwarf planet Pluto are in that sky! But keep in mind, it is still twilight (the sky is rendered bluish – in the eBook version - and the Sun is visible above the horizon), so we would have to wait until the Sun fully sets and it is dark enough for us to see the elusive Mercury. [The viewing time window for Mercury is small, for like a faithful puppy it sticks very close to its master, the Sun, and so as the Sun sets / rises, Mercury does the same, before or after, according to whether it is ahead or behind its master.]

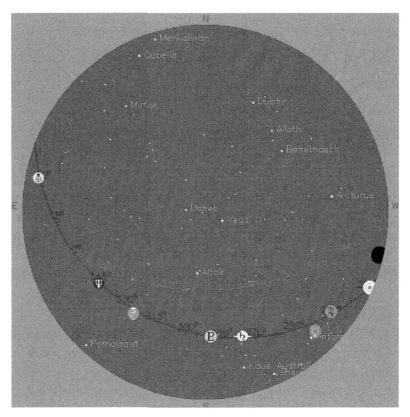

42 At Cornwall, Ontario, on Nov 6, 2018 at 4:30 pm (from "Your Sky" by Fourmilab)

Another bonus in the above image is that we can see that the locations of the planets is random in their distances apart. The order is random too. For example, Jupiter is the first planet from the Sun in the image whereas in reality it is 5th away from it. Similar comments apply to the other planets. For example, Pluto in reality is furthest from the Sun whereas in the image Uranus (7th furthest in reality) is seen as furthest from the Sun. This is so because as they circle the Sun they *appear* to come close to it and then move away, from Earth's viewpoint.

Now let's say you were looking heavenward and noticed this bright star in the sky, wondering what star it is. To identify it, you use a graphic rendering tool, such as Fourmilab's "Your Sky" (see **Appendix F: Night Sky Tool**, used for the above screen shots) showing the night sky for the date-time and location of your observation, then locate it visually in the vicinity of where you observed it.

This basically is the graphic approach: observe the sky, match it on a graphical display and identify. Although this may come across as simplistic, there are tools that allow you to take it up a notch or more. [Even Fourmilab's "Your Sky" has more features than seen above.] I used "Your Sky" for a couple of decades, making good use of it. More recently, I also use a powerful graphic tool called "Stellarium". The big difference between the two is that although both are graphic, the former is parameter-driven and web-based, the latter is GUI (graphical user interface) and a desktop app. For example, to change what stars are to be named, in "Your Sky" you type in a *Magnitude* in the "Names for magnitude" box and check off the box alongside, then click "Enter" to have it re-render, whereas in "Stellarium" you drag a slide bar and right before your eyes star names appear or disappear depending on whether you are dragging left (less names) or right (more names). A significant difference between parameter-driven and GUI is that the former requires you to update (refresh) the rendering every time a parameter(s) is / are changed, whereas in the latter, the rendering is updated on-the-fly (i.e. right there and then, in front of your eyes) whenever a GUI control (e.g. slide bar) is changed. These two examples reflect the availability of technological development tools and resources (i.e. state-of-the-art) in the late 1990s ("Your Sky") and the 21st century ("Stellarium"). More on Stellarium in **Appendix F: Night Sky Tool**.

Sketching Exercise

A useful exercise at this point is to sketch parts of the *CS* that show some of the interesting stars, constellations and asterisms in the southern hemisphere. Here is a suggested approach:

- Start with the blank sketch sheet below.
- Mark as an asterisk the approximate location of Sirius the Dog Star (brightest in the sky, in constellation Canis Major), label it "Sir".
- Mark Canopus (2nd, constellation Carina) label "Can", Rigel Kentaurus (3rd, constellation Centaurus), label "RiK", Procyon (8th, constellation Canis Minor) label "Pro".
- Mark Antares (16th) label "Ant". Draw and label its Zodiac constellation Scorpius (which we have done before so you may want to use it as a guide figure **34 Venus, Saturn, Jupiter**).

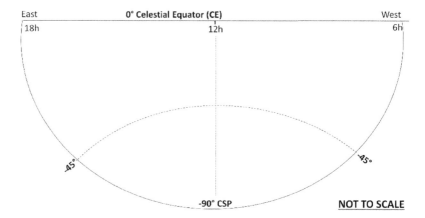

43 Blank sketch sheet

- Label the approximate centres of Zodiac constellations: Libra, Virgo and Leo (one of these sits at the **CE**). Approximate centres are given in **Table of All Constellations, their Stories** in **Appendix L: Information Tables.**

- Label the approximate centres of constellations: Crux, and the three parts of the retired constellation Argo Navis, viz. Vela, Puppis and Carina.

- Sketch an arc that represents the Milky Way band. Use **Table of Coordinates: by Glat-Glong** in **Appendix L: Information Tables** to look for Galactic Latitudes in the range ±5°, having Galactic Longitudes in the range 180° to 359.9° e.g. Shaula in constellation Scorpius has galactic coordinates -2°, 352°. Then for each such qualifying star, use the corresponding celestial coordinates to mark a dot through which the Milky Way band passes. This approach will only yield a few qualified stars (listed in the table below), not enough to plot the Milky Way band's path through our sketch area, so sketch its path for only the length you are able to. [Alternatively, you can use https://ned.ipac.caltech.edu/coordinate_calculator to produce precise points, shown in the table below as CalTech 1, 2, etc.]

Star	RA	Decl	Glat	Glong
Naos	8.1	-40	-5	256
Girtab	17.6	-39	-5	351
Shaula	17.6	-37	-2	352
Rigel Kentaurus	14.7	-61	-1	316
Acrux	12.4	-63	-0	300
Hadar	14.1	-60	1	312
Suhail/Regor	9.1	-43	3	266
δ Crux	12.3	-59	4	293
Mimosa	12.8	-60	3	303
CalTech 1	6.8	3	0	210
CalTech 2	9.2	-48	0	270
CalTech 3	17.3	-37	0	350

- Sketch a line that represents the Ecliptic. It's actually an arc but in the confines of the quarter sphere of your sketch, a straight line is a good approximation.
- Mark Voyager 1's current location (*RA17.2h, Decl12*) and label it "V1".

Use the tables in **Appendix L: Information Tables** for celestial coordinates for the above stars and constellations.

To check your work, match it against the following figure:

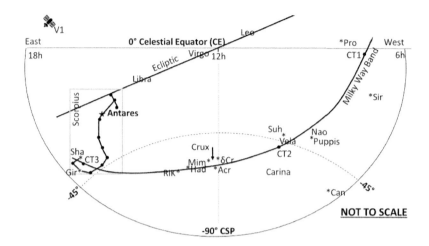

44 Sketch Solution

8. Star Naming

For everyday astronomy, most visible stars have names. [Some stars have two names e.g. the handle end of the Big Dipper/ladle asterism of Ursa Major is called Alkaid but also Benetnasch. Also, there is the odd case of two stars named Markab and Markeb, in constellations Pegasus and Vela, respectively!] In addition to that, these stars (many others visible through telescopes) have conventional names consisting of the name of the constellation a star belongs to, preceded by a Greek letter whose alphabetic order indicates the star's brightness rank within the constellation e.g. **Vega** is the common name of a star whose conventional name is **α** (alpha) **Lyra**. Alpha being the first letter in the Greek alphabet means Vega is the brightest in constellation **Lyra the Lyre**. But a word of caution, cartographers have been known for mislabelling and we are stuck with it e.g. the most glaring (sorry for the pun) example being **Sagittarius the Centaur**'s brightest star **Kaus Australis** which is conventionally labelled **ε** (epsilon) **Sagittarii** which would mean it is its 5th brightest! [Note the use of the genitive Sagittarii, which then translates as the epsilon of Sagittarius. But, **ε** (epsilon) **Sagittarius** is also acceptable.]

The above conventional naming was the work of Johann Bayer, a 17th century German astronomer. But it being limited to designating up to 24 stars per constellation (24 being the number Greek letters), John Flamsteed, an 18th century English astronomer (the first Astronomer Royal) came up with a numbering system, numbers assigned from West to East (i.e. in increasing order of **RA**s). Thus, star Vega is known in one of three ways:

Vega (common name)

α (alpha) **Lyra** (Bayer designation)

3 Lyra (Flamsteed designation)

There are also designations for deep-sky objects (galaxies, nebulae, star clusters): Messier, NGC (New General Catalogue). These do not concern us for our naked eye astronomy.

Tip: Many star names begin with "Al" (also, a few with "El") e.g Altair of the Summer Triangle asterism. The **Table of Coordinates: by Common Name** in **Appendix L: Information Tables** lists some 40 such names. This is not a coincidence, as "Al" or "El" are Arabic definite articles. Many star names are Arabic, just as many others are Greek or Latin.

Greek Letters
Here are Greek letters in alphabetic order: (left to right, down each row)

α	β	γ	δ	ε	ζ	η
alpha	beta	gamma	delta	epsilon	zeta	eta
θ	ι	κ	λ	μ	ν	ξ
theta	iota	kappa	lambda	mu	nu	xi
ο	π	ρ	σ	τ	υ	φ
omicron	pi	rho	sigma	tau	upsilon	phi
χ	ψ	ω				
chi	psi	omega				

Old versus Modern Constellations
The ancient astronomers identified all stars visible to them, organizing them into some 48 constellations. But with the advent of and improvement in viewing technology, a lot more stars were identified, many outside, many inside the old constellations. The IAU (International Astronomy Union) undertook to prepare a set of constellations to completely cover the entire sky, retaining the old ones, and adding new ones to fill gaps. For our naked eye astronomy, the old constellations are our mainstay as they contain nearly all the stars visible to the naked eye.

A useful exercise is to study the stick figure of a constellation and its stars, then draw from memory the stick figure, naming some of its brightest stars, by name and by Bayer and Flamsteed designations. Orion is suggested, but choose whichever that appeals to you.

9. Zodiac Constellations

Astronomy is what we are learning herein. But there is another similar term: astrology. The difference is, astronomy is concerned with mapping in 3D space, objects in space, and describing them. As compared to this, astrology is the practice of divining using planetary motions, just like other practices of divining using tarot cards, tea leaves, bones, beans, palm lines, etc. [Some, not shy about it, call all this, very bluntly: quackery, voodoo, ...] So, astrology is to be ignored by astronomers? In principle, yes. But, it is important to know some of their terminology, so that when encountered we know where it came from. [Another related science (yes, science, not quackery) is cosmology which is about the history of the universe, mainly what is its origin, its nature, and where it's headed. One of the most famous cosmologists was Stephen Hawking 1942-2018.]

What is of astrological interest to us amateur astronomers is the term "zodiac" (meaning a circular course). Zodiac refers to the set of constellations that are along the path of the Ecliptic (the apparent path of the Sun upon the celestial background). The Ecliptic cuts through 12 constellations viz. Aries, Taurus, Gemini, etc. Well, ancient astrologers went to 'work' on this and to 'divine' the future (the less shy would say hoodwink the gullible) by associating them with calendar date ranges, collectively making up 12 Zodiac months, and associating these ranges with birthdates of clients e.g. Capricorn's date range is December 22 to January 20. Thus, someone born January 1st is a Capricorn, and is therefore relentless like its mountain goat! How did the ancient astrologers (yes, they have a long history, as long as witchcraft's) assign these date ranges? By the then transit times of the Sun into and out of Zodiac constellations. But that was then. Now, after centuries / millennia, these transit times have shifted by a month (due to Earth's wobble called Precession, over a cyclical period of 26,000 years). Thus, Capricorn dates today are January 19 to

February 15, which is no longer the old astrological Capricorn dates that astrologers continue to cling to, but which in reality are now the transit dates of Aquarius! [Jokers do overtime on this: An Aquarius went to see an astrologer and came back a goat called Capricorn!] Note that two of the Zodiacs are called differently by astrologers as compared to their astronomical names: Scorpio and Capricorn in astrology versus **Scorpius** and **Capricornus** in astronomy. Herein of course we use latter names.

Back to seriousness. A Zodiac constellation of interest to us is **Sagittarius the Centaur**, because looking at its right edge is looking at the centre of our **Milky Way** galaxy. Its γ (gamma) star, its 4th brightest, has the following Galactic Coordinates:

Glongitude: 1° [that is, one degree to the left of galactic centre]

Glatitude: -5° [that is, five degrees below galactic plane]

Sagittarius is shown in the figure below with its iconic **Teapot** asterism. The teapot's spout is near the **Galactic Centre**.

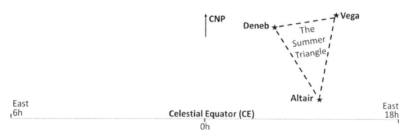

NOT TO SCALE

45 Sagittarius

Similarly, look at Zodiac constellation **Taurus the Bull** in the night sky and you are looking in the opposite direction

9 Zodiac Constellations   97 | P a g e

towards the edge of the galaxy, which is in the direction of the 180° line, the Galactic Anti-centre. Again, look at West of star **Deneb** in constellation **Cygnus the Swan** and you are looking in the direction of the 90° line, and in the opposite direction the 270° line.

10. Walkabouts in the Night Sky

The purpose here is to first trace out *the* main reference line in the sky – the **CE**. The **CE** is 'visible' from anywhere on Earth, even from the **CNP** or the **CSP** ('visible' in quotes because it is an imaginary circle). More importantly, it is readily located by an observer as we will see below. We also locate, perpendicular to the **CE**, a moving line (albeit slowly to the human eye), the **LM**. Then using it, we locate constellations / stars and point out along the way, the Milky Way band, the Ecliptic and the Zodiacs. We don't cover all 88 constellations here, only the well-known, containing the brightest stars which will be our focus for each such constellation. In addition to all that, we locate two other, most important reference points: the **CNP** and **CSP**. [A word on the usefulness of constellations in locating its stars and neighbouring constellations. Constellations are defined by their boundaries but are more popularly picturized by their stick figures. However, neither linework – boundaries nor stick figures – are there in the sky, and so neither are useful in locating constellations. So, one approach to finding a constellation is to locate its brightest star, by using its *Azimuth-Altitude* (see **Appendix B: Converting Celestial Coordinates to Local Horizon Coordinates**). From this star, the rest of the stars in the stick figure can be located. But better still, locate the brightest TWO stars so that you have a solid scale orientation of the stick figure.]

We do all this with only minimum, or no, calculations to find objects in the sky. To do this, the four cardinal points on the horizon have to be approximately known. Sunrise (the easterly horizon generally) and sunset (the westerly horizon generally) help to locate where North and South are (halfway between sunrise and sunset). North or South are more precisely located as follows, whichever is closer to your Latitude:

Find North or South

CNP is located by finding the Big Dipper/ladle asterism. It lies at **Declination** 50° to 60° (average 55°). Work out how far this average circle is from your **Zenith** e.g. if you are at Latitude 25°, this average **Declination** circle will be 55 – 25 = 30° away towards North, which for now is halfway between the local sunrise and sunset along the horizon. Look for the Big Dipper anywhere on this circle of radius (90 – 55) = 35°. It continually moves along this circle around the **CNP**. The diagram below describes how to locate the **CNP** once the Big Dipper is found:

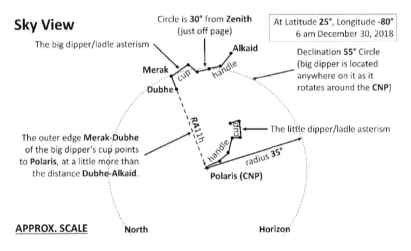

46 Finding North

CSP is located by finding the **Crux** constellation. It lies at an average **Declination** -60°. Work out how far this average circle is from your **Zenith** e.g. if you are at Latitude -50°, this average **Declination** circle will be (60 – 50) = 10° away towards South, which for now is halfway between the local sunrise and sunset along the horizon. Look for the Crux anywhere on this circle of radius (90 – 60) = 30°. It continually moves along this circle around the **CSP**. The diagram below describes how to locate the **CSP** once the Crux is found: [There are bright stars galore surrounding the **CSP** to help in its identification. Not surprisingly, the Milky Way band cuts right through the Crux constellation, as well as the Centaurus constellation of which the brightest stars are Rigel Kentaurus and Hadar.]

Sky View

At Latitude -50°, Longitude -60°
December 30, 2018
6 am

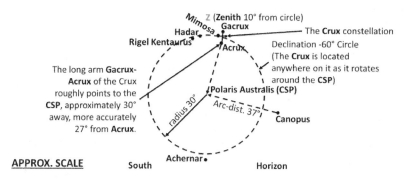

47 Finding South

Arc-distances to **CSP** from the surrounding bright stars can serve to fix its location e.g. Canopus (2nd brightest) in the above figure. See **Appendix G: Cosine Rule of Spherical Triangles: Arc-distance Calculation** for how to calculate arc-distances.

Tip: The Big Dipper is visible from anywhere in the northern hemisphere, ditto the Crux in the southern ('visible' in quotes to indicate that although these are up there in the sky, they are truly visible at night only). From about Latitude 40°N and the **CNP**, the Big Dipper is *Circumpolar*, and from about Latitude 35°S and the **CSP**, the Crux is *Circumpolar*, meaning these asterisms are up in the sky all night long, never rising nor setting, thus no waiting is required for them to show up (rise). But, between each of the above Latitudes and the Equator, one has to wait for the asterisms to rise :-(The easiest way to find out what the viewing time window for one of these asterisms is to use a tool like Stellarium, making it spin the sky through the night and noting the rising and setting of the asterism – see **Appendix F: Night Sky Tool**.

Using *RA* Lines at *CNP* or *CSP*

To have a mental image of ***RA/Declination*** grid lines at your observation point, the following analogy of an umbrella might be useful:

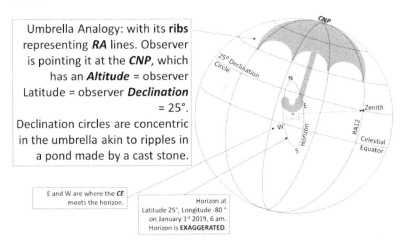

Umbrella Analogy: with its **ribs** representing ***RA*** lines. Observer is pointing it at the ***CNP***, which has an ***Altitude*** = observer Latitude = observer ***Declination*** = 25°. Declination circles are concentric in the umbrella akin to ripples in a pond made by a cast stone.

E and W are where the ***CE*** meets the horizon.

Horizon at Latitude 25°, Longitude -80 ° on January 1ˢᵗ 2019, 6 am. Horizon is **EXAGGERATED**.

48 Umbrella Analogy for Grid Lines

In this analogy, the observer is facing North, pointing an open umbrella at the ***CNP*** (or facing South and pointing to the ***CSP***) in which the ***RA/Declination*** grid lines can be imagined. This orientation aids in measuring out degrees along these imaginary lines in the sky.

But we do something very useful with the centre of the above grid as shown in the following diagram:

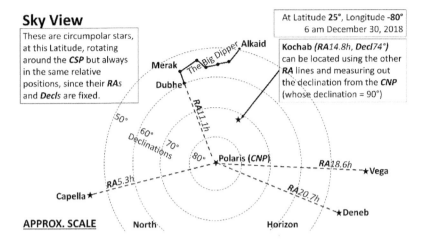

Sky View

These are circumpolar stars, at this Latitude, rotating around the *CSP* but always in the same relative positions, since their *RAs* and *Decls* are fixed.

At Latitude 25°, Longitude -80°
6 am December 30, 2018

Kochab *(RA14.8h, Decl74°)* can be located using the other *RA* lines and measuring out the declination from the *CNP* (whose declination = 90°)

Alkaid

Big Dipper

Merak

The Big Dipper

Dubhe

RA11.1h

Kochab

50°

60°

Declinations

70°

80°

Polaris (CNP)

RA18.6h

★Vega

RA5.3h

RA20.7h

Capella★

★Deneb

APPROX. SCALE

North

Horizon

49 RA Lines at CNP

The stars in the diagram are not difficult to locate, especially the Big Dipper and having found it, not only is the *CNP* located but the three bright stars (Vega the 5ᵗʰ brightest, Capella the 6ᵗʰ, Deneb 19ᵗʰ) can then be located by sight or by using the Dubhe-Polaris *RA11.1h* as a reference and measuring out the **Declination**. Once these three bright stars are located, we now have a framework to easily locate other stars in the vicinity as seen in the above diagram showing how Kochab is found. (Kochab is part of the Little Dipper asterism, located on it at the same relative location as Dubhe is in its Big Dipper asterism).

RAs emanate from the celestial poles (recall the above umbrella analogy) and we can put this knowledge to use in locating nearby stars. For example, the line Polaris-Dubhe is *RA11.1h*, and similarly the other *RA* lines shown in the above figure. To locate Kochab (*RA14.8h, Decl74°*) we would need to sweep out an angle of 3.6h (14.7h – 11.1h) at the *CNP* from the Polaris-Dubhe line. 3.6h = 3.6 X 15 = 54°. But sweeping out degrees is not as easy as measuring arc-distances by fist-n-thumb. Right-angles are relatively easier to sweep out than other angles, so we sweep out just under two-thirds of a right angle, and then lay out 26° (90 − *Decl74°* of Kochab) using fist-n-thumb! Sweeping out angles being subjective, this technique requires much practice and experience. Alternatively, Kochab's *RA14.8h* being the average of the Polaris-Dubhe's *RA11.1h* and Polaris-Vega's *RA18.6h*, the mid-line between the latter two puts you in the right direction.

In general, the above method requiring eyeballing the *RA* direction can lead to inaccuracies over a long distance from the *CNP*. Thus, it should be restricted to a suggested maximum distance of 30°.

The same technique as above is applicable to the *CSP*. See the equivalent diagram below:

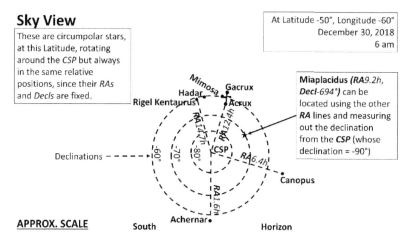

50 RA Lines at CSP

The example in the above figure of locating star Miaplacidus requires laying out a line between between *RAs* 6.4h and 12.4h and for an arc-distance of 21°, being the difference between its *Decl* and that of the *CSP*. The problem with the *CSP* is that it does not have a bright star marking its location, although it does have one, Polaris Australis, but which is quite dim.

Trace the *CE*

Facing the **CNP** (or the **CSP**), trace an arc from your **Zenith** to it, continuing on to the horizon: that establishes the North (or the South) point on the horizon. The point directly opposite to it on the other side on the horizon will be the South (or the North) point. If your Latitude is positive (meaning you are North of Earth's Equator), face the South point of the horizon, and if negative, face northward. Finally, trace an arc from your **Zenith** for a distance equal to the absolute value of your Latitude. That's where the **CE** passes through.

For the steps from here on, you would need to carry a list of at least the brightest (1ˢᵗ Order **Magnitude**) 21 stars with their **RA**s and **Decl**s. See **Table of Coordinates: by Brightness** in **Appendix L: Information Tables.**

Local Sidereal Time

LSidT = **RA** of **LM**: e.g. say LSidT is 4h 30m (obtained from a smart phone app, for example). Then, at this exact moment, the arc from the **CNP** (or the **CSP**) through your **Zenith** goes on to cut the **CE** at **RA4.5h**. Thus, the rear shoulder of **Orion the Hunter**, being the star Betelgeuse with **RA5.9h** will be about an hour and a half (= 22.5°, 1 hour = 15°) East of it, at that point in time, meaning our **LM** will begin sweeping through the hunter in about that time, as it inches towards it (it actually is the Earth which is rotating, making it appear that the sky and its stars are rotating).

To use the LSidT from here on, you would need to note the difference between it and your watch. Then, you can tell what the LSidT at any instance will be, by applying this difference to your watch time.

CE Constellations & Stars

There are many constellations / stars that straddle or are near the **CE**. Any time during the night, anywhere on Earth, one or likely more than one will be in the sky sometime during the night. **Orion the Hunter** could be one. It is one of the easily recognized constellations. We can find it in the sky just by looking for it on the *CE*, or using the LSidT as above, or we can locate it using the local *Azimuth* and *Altitude* of the top of his belt stars which is just below the *CE*, appearing to touch it. His two bright stars, Rigel the 7th brightest and Betelgeuse the 10th, straddle the *CE*, 8° South and 7° North of it, respectively. [Incidentally, the Milky Way band cuts the *CE* just East of Orion.]

We now have located two points on the *CE*: where it cuts our *LM* and the top of the hunter's belt.

Again, we can use the LSidT to locate Altair, 12th brightest, in constellation **Aquila the Eagle**, with an *RA19.8h* in the same way as above, its *Declination* telling us it is 9° above the *CE*, being about as much above it as Rigel is below. Or we could use its local *Azimuth* and *Altitude* to locate it. Note that the eagle's southern wing just cuts the *CE*, with the wing tip 0.8° below it – see figure **23 The Summer Triangle**. This wing tip is a good distance West of the hunter, about 80% of the length of the visible *CE*.

With now three points on the *CE*, we have a good idea of its entire path in the local sky, noting that where it cuts the horizon are the cardinal points East and West. Voila – the whole *CE* is now fixed in the sky!

Tip: Think of a star's *RA* as its 'timestamp' (i.e. *Sidereal Timestamp*), meaning it is fixed (permanent) just like a timestamp. Then, to continue this analogy, your 'wristwatch' (i.e. your smartphone *Sidereal Time* app) tells you the LSidT. Both, the timestamp and the wristwatch, keep to *Sidereal Time*, with the difference that the timestamp is permanent while the wristwatch is ticking (your LSidT is always ticking). Now, if a timestamp is *RA19.8h* (which is star Altair's *RA*) and your wristwatch says the LSidT is 15.0h, then Altair will arrive / transit in 4.8h (19.8 – 15.0).

The LSidT and your Latitude (which equals to the *Declination* circle overhead) combine to specify your *Zenith*, which can be plotted on our *CE* model to know which dome / portion of the *CE* is in the field-of-view e.g. if the LSidT at Collingwood, ON (Latitude 45°N) is say 21h, then these coordinates equate to, at that instance, celestial coordinates of *RA21h* and *Decl+45°* which, when plotted on the *CE* model, is quite close to star Deneb, and if you imagine yourself at the centre of the *CE* model (which is Earth) and you point a metaphoric open umbrella (figure **48 Umbrella Analogy for Grid Lines**) to Deneb, it tells you the current dome of the *CE* that is your field-of-view!

Tip: Any of the stars close to the *CE* also serve as *RA* markers e.g. star Mintaka, the top of Orion's belt, just touches the *CE* at *RA5.5h*. That is, Mintaka serves as the marker *RA5.5h* on the *CE*, from which other markers can be laid out to assist in finding other objects.

Another bright star near to the *CE* is **Procyon the Little Dog Star**, 8th brightest, in constellation **Canis Minor**. It is about 5° North of it, about 30° East of the hunter as measured along the *CE*. Its brighter kin, **Sirius the Dog Star**, the brightest, in constellation **Canis Major**, is nearby. Use its local *Azimuth* and *Altitude* to find it.

Another bright star is in the vicinity: **Aldebaran**, 14th brightest, in constellation **Taurus**, a Zodiac constellation, thus the Ecliptic cuts through it. And a bonus: by coincidence, the Milky Way band passes through it – see figure **22 Celestial Sphere showing the Milky Way band**.

Declination Stars & Constellations

Your Latitude = Your *Zenith*'s Declination: The declination circle that runs through your Zenith has stars of that declination (and other close ones) e.g. the declination circle overhead of Washington, DC (Latitude 39°) contains Vega (*Decl39°*), the 5th brightest star, as well as constellation Cygnus the Swan (average *Decl40°*), etc. We use this knowledge to continue with our walkabout:

Using the two Latitudes, 25° and -50°, used above in locating **CNP** and **CSP** as examples, we explore what we can find there. The **Table of Coordinates: by Decl-RA** in **Appendix L: Information Tables** tells us that in the vicinity of *Declination* circle 25°, the nearest bright star is Gemini the Twins star **Pollux** (17th brightest), which is just 3° above it and very close to the **RA** of **Procyon** (previously located above). Less than 5° away is his half-brother **Castor**, can't miss it. Gemini being a Zodiac constellation, it's on the Ecliptic path. Previously above, we located star Aldebaran in Zodiac constellation Taurus, an immediate neighbour of Gemini. So, we have a segment of the Ecliptic path, namely Taurus-Gemini. [Incidentally, the Milky Way band passes right between these two Zodiac constellations.] If we extend this segment from Gemini further down the Ecliptic path, we would enter Zodiac constellation Cancer. Unfortunately, its brightest star just barely makes the ranks of the 200 brightest! So, we continue beyond it along the path to the next Zodiac, Leo, whose brightest star Regulus (21st brightest overall) makes the ranks of First-order *Magnitude* stars. Its distance from Pollux is about 36° which can be laid out using fist-n-thumb. Continuing on the path, when we cross the **CE**, we are in the middle of the next Zodiac, Virgo, containing star Spica (15th brightest overall). The takeaway from all this is that you can explore along well-known great circles e.g. the **CE**, the Ecliptic, the **LM**, selected **RA**s.

As for Latitude -50°, the **Table of Coordinates: by Decl-RA** in **Appendix L: Information Tables** tells us that in the vicinity of *Declination* circle -50°, the nearest bright star is **Canopus** (2nd brightest), which is less than 3° below it and very close to the **RA** of **Sirius**, the brightest star. Within a dozen degrees of *Declination* circle -50°, are a horde of bright stars viz. Acrux (13th brightest and part of the nice little – the smallest in fact – constellation called the Crux (a.k.a. the Southern Cross), Rigel Kentaurus (3rd), Hadar (11th), Mimosa (20th), Achernar (10th).

Local Meridian-RA Stars & Constellations

Say, we are going to observe tonight at 9 pm, and that the LSidT at that time is 6:15 (= 6.25h), which represents the **LM** at that exact time and that some star with **RA6.25h** will be transiting. [See **Appendix A: Obtaining Local Sidereal Time**.] The **Table of Coordinates: by RA-Decl** in **Appendix L: Information Tables** tells us that the nearest First Order **Magnitude** star to this **RA** is Canopus (2nd brightest), with **RA6.4h**, meaning that 0.15h (6.4 – 6.25) after 9 pm, Canopus will transit. Its **Decl-53°** determines how high up in the sky it will be then, based on what your Latitude is. This is easily done using the fact that your Latitude value is the same as the **Declination** circle at your **Zenith**. The angular distance between the **Zenith** and the star's **Decl-53°** is the complementary angle of the star's local **Altitude** e.g. if you are at Latitude 20°N, you subtract this angular distance from 90° to obtain the **Altitude**: 90 – (20 + 53) = 17°. If you were at Latitude 20°S, **Altitude** = 90 – (53 – 20) = 57°. But, before you can use this for your observation, you need to check if this star is in your blackout zone (discussed in **Preparations for Finding Orion** in **Appendix K: Preparing for Observation**), which is a circular zone around the celestial pole furthest from you, with a radius that is equal to your Latitude. Thus, for the above example of Latitude 20°N, all stars within a circle around the **CSP** of radius 20° (= **Decl-70°**) will not be visible, so Canopus' **Decl-53°** tells us it will be visible. As for the other example, it being *South* of the Equator, the blackout will be around the **CNP** and so Canopus' **Declination** puts it obviously outside it. Ergo, it will be visible in both cases and you can look for it on your **LM** at its calculated **Altitude**.

Now for some more goodies! A scan of the **Table of Coordinates: by RA-Decl** in **Appendix L: Information Tables** tells us that the nearest First Order **Magnitude** star to Canopus is Sirius (the brightest) with **RA6.8h**. Thus, less than half an hour (6.8 – 6.4 = 0.4h) after we have observed Canopus transiting, Sirius is ready to do the same. The next nearest First Order **Magnitude** star to Canopus is Betelgeuse (8th brightest) with **RA5.9h**, half an hour (5.9 – 6.4 = -0.5h) *before* we have observed Canopus transiting, Betelgeuse will have done the same. Not a bad haul, three of the brightest stars in the span of less than one hour! But, we are not done with the goodies, for in between these three, the above table tells us there are four other stars of which three are ranked in the top 50 in brightness viz. Menkalinan (40th), Alhena (42nd) and Mirzam (46th). Moreover, you can fill your goodie bag with more stars to observe by extending your observation time window!

Note that the above approach is based on a star's transition, but we know that it is will be visible for a considerable time *before and after* transition. Why the focus on transition then? Because it happens on your **Local Meridian** which is readily known, being the **Zenith** North-South great circle: you stand facing North or South, as a result of which you are directly facing the Northern or Southern arc of the **Zenith** circle and can watch the stars you are interested in, to approach, transit through and move away from your **Local Meridian**.

The following is a format of the notes you could make for the above example:

Date: Wednesday, February 13, 2019. Lat20°N, Long80°W.
Time: 9 pm. LSidT: 6:15 (= 6.25h).

Star	RA	Decl	Alt	Az*	9 pm
Betelgeuse	5.9	7	77	180	-0.35h
Menkalinan	6.0	45	65	0	-0.25h
Canopus	6.4	-53	17	180	+0.15h
Mirzam	6.4	-18	52	180	+0.15h
Alhena	6.6	16	86	180	+0.35h

| Sirius | 6.8 | -17 | 53 | 180 | +0.55h |

Azimuth is either 0 (North) or 180 (South), according to whether a star's **Declination** is North or South of your Latitude.

Bright Star Regions

There are two regions rich in bright stars, radiating out from: (1) Betelgeuse in constellation Orion, and (2) the CSP. (1) is in the diagram below:

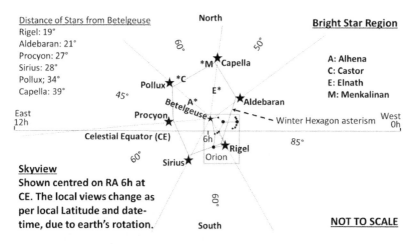

51 Bright Star Region

It has seven of the total 21 First-order *Magnitude* stars. (2) is seen in figure **47 Finding South**, and is self-explanatory. It has six of the total 21 First-order *Magnitude* stars, thus between it and the above region, they have 15 of the total 21 First-order *Magnitude* stars, within their average radii of 30°. The first region though, is visible from almost anywhere on Earth, as the **CE** cuts through it, whereas the latter region is a blackout zone for Latitudes above 30°N.

11. Light Pollution Buster

One way of overcoming light pollution is taking long-exposure photos and identifying at leisure the stars and their constellations. I have enjoyed doing this, using nothing more than a pocket digital camera. It's like playing 'detective'! Let's see how:

Observations at Hollywood, FL, mid-January, 2019

Below is a photo I took with my Canon Power Shot Elph 110 HS. I have circled some of the relatively more visible stars.

Tip: How to mark, label, etc observation photos as seen below? I use MS PowerPoint, one slide per photo. I copy and paste the photo and then look for tiny fuzzy grey dots and draw a circle around each, as seen below.

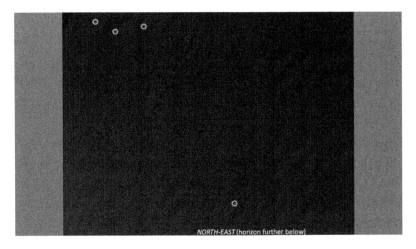

52 Hollywood, mid-Jan 2019 – 1

The photo has NOT been doctored in any way, which has its pros and cons: the main pro is that the stars are seen realistically. The main con is that they are hard to see; they are best viewed with lights turned off. Zooming in (or using a magnifying glass) too makes them more visible. **Tip**: [in MS PowerPoint] Any tiny greyish object is better seen if it is in motion. If you scroll the photo, horizontally or vertically, any tiny fuzzy grey dot in it will scroll too, while a dirt mark on your screen will remain stationary. Voila!

The next step is to use a tool to render the night sky of this area. This photo was taken in the north-easterly direction, with its bottom border about 30-40° above the horizon. Here is a screenshot of Stellarium's rendering of the area. [Stellarium is described in **Appendix F: Night Sky Tool**.] With a bit of trial-n-error matching, the four stars pointed out in the above figure are identified as Vega (near the middle-bottom), and the three near the top-left corner as Alioth and Alkaid – the middle star is shown in the screenshot but not labelled, as the focus at this stage is the brightest stars in the area.

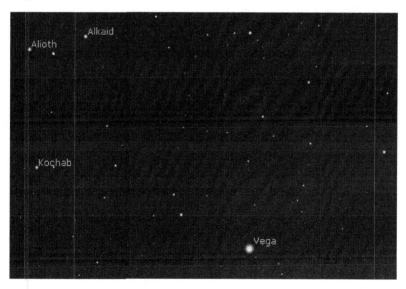

53 Hollywood, mid-Jan 2019 – Stellarium screenshot

Thus, we can label these three stars, as seen below:

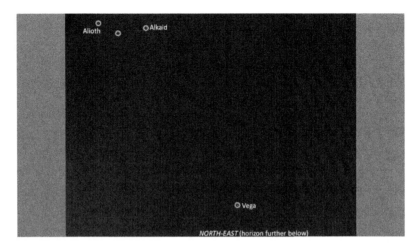

54 Hollywood, mid-Jan 2019 – 2

[Alternatively, or in combination with the above method, I could have field-measured the **Azimuth-Altitude** of a few stars, identified them and used them in orienting the photo to a rendered sky.]

Now we are ready to roll, as having anchored the photo to the corresponding rendering in Stellarium, it's easy to identify the other stars and a lot of fun and satisfaction. The names of unlabelled stars are easily obtained by clicking on them in Stellarium. Below is the result:

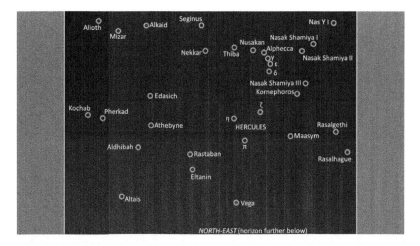

55 Hollywood, mid-Jan 2019 – 3

The big takeaway here is that, although in the field I could readily see only a fraction of the stars in the above photo, back in the 'lab' (read MS PowerPoint), I could zoom in and see 30+ stars! That's why I call this Light Pollution Buster :-)

And now, the *pièce de résistance* – constellations! Stellarium can render these, as guides for us to draw constellation linework and labels:

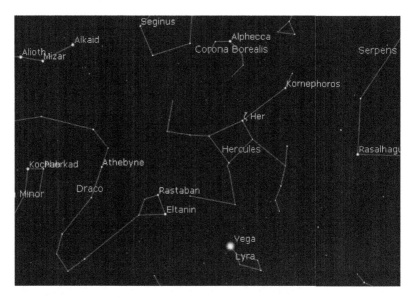

56 Hollywood, mid-Jan 2019 – Stellarium screenshot 2

Using the above as a guide, we update our photo:

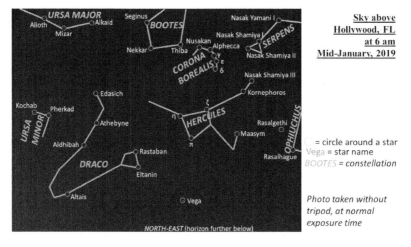

57 Hollywood, mid-Jan 2019 – 4

It is a good idea to throw in a few grid lines, for orientation:

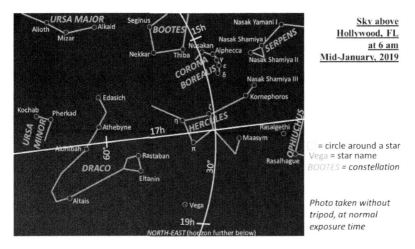

58 Hollywood, mid-Jan 2019 – 5

This technique was used in the following observations which show only the end result, omitting the above step-by-step process, plus commentary. So that these photos can be related to the **CS**, the centres of the photo coverage for each site have been marked on the **CS** model that was previously constructed, as seen below: [Look for four labels at the periphery of the **CS**: Ho=Hollywood, Co=Collingwood, Va=Varadero, Za=Zanzibar, with arrows pointing to their photo locations on the **CS**]

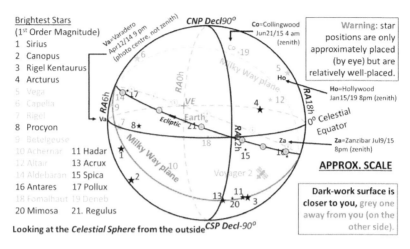

59 CS showing Observation Location Zeniths

Observations at Zanzibar Island, Tanzania, early July 2015

The fabled Zanzibar archipelago is located some 6° South of the Equator and some 38° East of Greenwich.

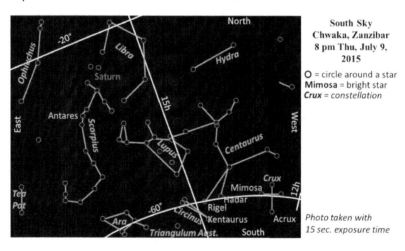

South Sky
Chwaka, Zanzibar
8 pm Thu, July 9, 2015

O = circle around a star
Mimosa = bright star
Crux = *constellation*

Photo taken with 15 sec. exposure time

60 Zanzibar Observation Photo 1

This was a good haul: the cute smallest constellation, the **Southern Cross (Crux)** (which in most of North America can't be seen by eye because it's just above the horizon at best), 10 other constellations and – **Saturn**. This was at the Eastern shore of Zanzibar Island, specifically Chwaka Bay, exactly on the opposite side of the island from touristy Stone Town of Zanzibar on the West shore. Three Zodiacs can be seen viz. **Libra the Scales, Scorpius the Scorpion** and the **Teapot** asterism which is inside **Sagittarius the Archer.** Five of the brightest (1st Order *Magnitude*) 21 stars are seen in this photo viz. **Rigel Kentaurus** (3rd brightest), **Hadar** (11th), **Acrux** (13th), **Antares** (16th) and **Mimosa** (20th), the bottom-right corner being one of two regions on the *CS* having a good collection of the brightest stars (together the two regions having 13 of the 21 brightest stars). Here are some more photos at the same location: [There are more stars in these photos than have been circled. I had to resist the urge to circle them all, in order to keep clutter down!]

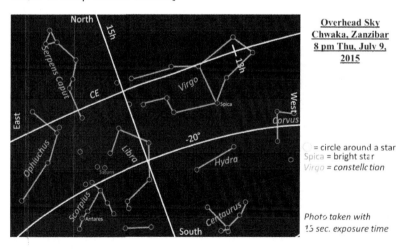

61 Zanzibar Observation Photo 2

The bottom half of the above photo roughly covers the same area as the upper half of the previous. One more Zodiac is seen: **Virgo**

the Virgin, with its bright star Spica (15th). Zanzibar Island's light pollution was not bad, especially at the Eastern shore.

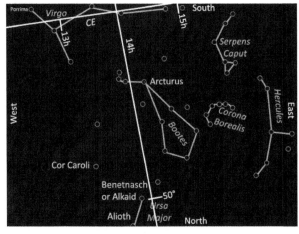

Another 1st Order *Magnitude* star, **Antares** (16th brightest).

62 Zanzibar Observation Photo 3

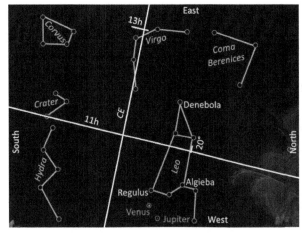

63 Zanzibar Observation Photo 4

Another Zodiac, **Leo the Lion** containing a 1st Order *Magnitude* star, **Regulus** (21st brightest), plus two more planets: Venus (the brightest) and Jupiter (the 2nd brightest).

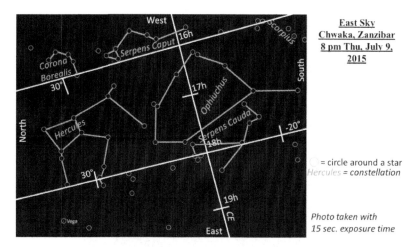

64 Zanzibar Observation Photo 5

Another 1st Order Magnitude star, **Vega** (5th brightest).

The above photo's focus is approximately the **CE** at **RA17h**, in the polygonal constellation Ophiuchus, which is background to the current location of NASA probe Voyager 1, and which separates the Serpens constellation into two, Serpens Caput (North of **CE**) and Serpens Cauda (South).

Observations at Collingwood, Ontario, Canada late June 21, 2015

Collingwood, ON competes with Huntsville, ON for the title of Ontario's Vacation Capital. It certainly was the Ship-building Capital of Ontario until the mid-1980s when its last one was launched into the Georgian Bay. It is located some 45° North of the Equator and some 80° West of Greenwich.

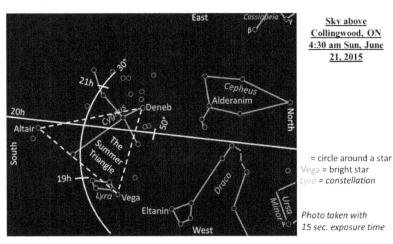

65 Collingwood Observation Photo with the Summer Triangle

The most interesting item in the above photo is the Summer Triangle asterism whose three corner stars viz. **Vega** (5th brightest), **Altair** (12th) and **Deneb** (19th) belong to different constellations, respectively **Lyra the Lyre**, **Aquila the Eagle** and **Cygnus the Swan**. Interestingly, Vega was visible both in Collingwood here and in Zanzibar above, which figure **59 CS showing Observation Location Zeniths** above shows that their sky views overlap, as does the one in Hollywood. Varadero below does not have Vega in sight at its observation time (overnight though it would have come into sight).

Being centered at a high **Declination**, no Zodiacs nor planets are visible in this photo. But, another photo I took (below) has a couple of interesting things.

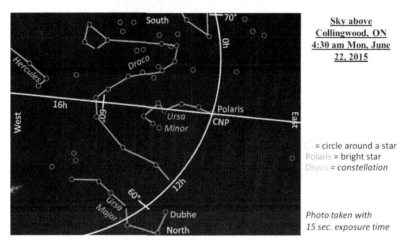

66 Collingwood Observation Photo with Polaris

The above photo clearly shows how the outer edge of the Big Dipper asterism (which is part of **Ursa Major the Big Bear** constellation) points to Polaris (which itself is part of the Little Dipper asterism, being part of the **Ursa Minor the Little Bear** constellation). Also seen in the photo is constellation **Draco the Dragon**, showing its head-to-tail skeleton. Some more constellations are seen in the next photo, below:

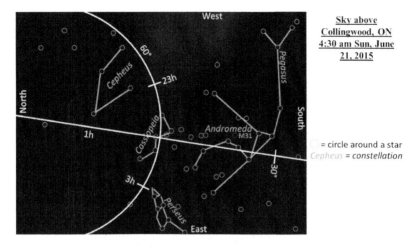

67 Collingwood Observation Photo with Cassiopeia

The **King & Queen of Ethiopia, Cepheus & Cassiopeia**, respectively, and their **Princess Andromeda** are centre-stage in this photo. **M31** in constellation **Andromeda the Chained Woman** (chained by her king-father to appease sea monster Cepheus) is the furthest of three galaxies visible to the eye, but it is hard to see, yet it's there, when you take at a highly-zoomed-in look.

Observations at Varadero, Cuba late September 2014

Varadero is Cuba's premier resort area, resorts fringing the northern shores of the Varadero peninsula. It is located just below the Tropic of Cancer (some 23° North of the Equator) and some 81° West of Greenwich.

Below is an auto-exposure photo taken without a tripod and it is much clearer than the above photos, indicating low light pollution:

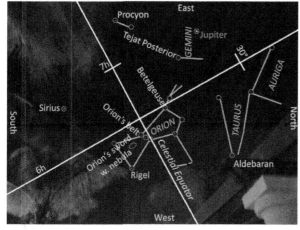

○ = circle around a star
Sirius = bright star
ORION = *constellation*

Photo taken without tripod, at normal exposure time

68 Varadero Observation Photo

It seems to have 'everything' in it: the **CE**, the Ecliptic (represented here by Zodiacs **Gemini the Twins** and **Taurus the Bull** and planet **Jupiter**), five of the brightest 21 stars viz. **Sirius** (brightest), **Rigel** (7th), **Procyon** (8th), **Betelgeuse** (9th) and **Aldebaran** (14th). Not only are Orion the Hunter's bright stars easily seen but also in the middle of his sword is the **Orion Nebula**, clearly visible!

Of the 21 brightest stars, 16 were photographed at the above four observation sites!

Glossary

Altitude (Alt): This is one of the Local Horizon Coordinates. From any point "X" on the horizon, trace an arc upwards in the sky, directly to your overhead (call it point "Z" for *Zenith*). We have just traced an arc of 90° from X to Z. If we had traced it only half-way up to the "Z" it would have been an arc of 45°. These are degrees of *Altitude* (which confusingly is the same term for height above mean sea level e.g. Kilimanjaro's altitude is nearly 20,000 feet above mean sea level). To obtain the *Altitude* of an object O in the sky, start at Z and trace an arc directly to O, continuing the arc on to a point (call it "Y") on the horizon. O's *Altitude* is the arc angle from Y to O, which you can measure using fist-n-thumb. See figure **9 Local Horizon Coordinates**.

Apparent Solar Time (ASolT): This is time as observed on a sundial, meaning it is the true local solar time e.g. a sundial will read 12 noon exactly when the sun is at its peak as it moves across the sky – at this point its gnomon casts the shortest, thinnest shadow of the day. It varies from the *MSolT* (which is what our normal clocks tell us and which represents the time over the entire local time zone) by about ±15 minutes, this varying difference being called the *Equation of Time*.

Axial Tilt: The earth's spinning axis is not perpendicular to the plane of its motion around the sun. It is tilted from the perpendicular by **23.5°**.

Azimuth (Az): This is one of the Local Horizon Coordinates. From your overhead (call it "Z" for *Zenith*) trace an arc directly to Polaris (see **Find North or South** in **10 Walkabouts in the Night Sky**) the North Star, continuing the arc on to a point (call it "N" for North) on the horizon. Next, for an object O in the sky, trace an arc from Z directly to O, continuing the arc on to a point (call it "Y") on the horizon. The arc angle along the horizon from N to Y is O's *Azimuth*, which you can measure using fist-n-thumb. See figure **9 Local Horizon Coordinates**. [A caution on what is horizon: Apparent horizon is the actual profile of the furthest topographical features e.g. if you are in a valley its apparent horizon will be at some *Altitude* higher than you. Conversely, if you are on the summit of a mountain, the apparent horizon will be at negative *Altitude*. True horizon on the other hand is the periphery of an imaginary horizontal plane coincident with the ground the observer is on. This is not difficult to imagine in the field. True horizon is very intuitive: hold out horizontally your arm in front of you with fore-finger pointing forward. Rotate yourself 360° and the fore-finger will have traced the true horizon :-) Another way to understand the distinction is, if the apparent horizon is jagged (e.g. mountain ranges) measurements along such a jagged profile would be longer than measurements along the true horizon. Yet another way to sink this in, is to imagine yourself at Times Square, NYC: you can't see any of the apparent horizon, for you are in the middle of Manhattan, surrounded by buildings. Yet, you can easily tell where true horizon is, as described above.]

Celestial Equator (CE): This is Earth's Equator projected on to the *CS*.

Celestial North/South Pole (CNP/CSP): These are points to which Earth's axis of rotation points, the northern end of this axis points to the *CNP*, the southern end *CSP*.

Celestial Sphere (CS): This is a model consisting of a vast sphere, on which all stars are imagined to be located. In reality they are *not* on a common sphere because they are at varying distances from earth, but being very, very far away from Earth, geometrically we can make the assumption that they are all on *one common* very large sphere.

Glossary

Circumpolar: A star is said to be ***Circumpolar*** if as it rotates in the sky about the ***CNP*** or the ***CSP*** it does not rise above nor set below the horizon. Thus, it always is in 'view' ('view' in quotes since the 24 hours it takes for it to complete its daily rotation includes daylight hours when it can't be seen, but it nevertheless is still above the horizon).

Declination (Decl), designated δ (delta) in formulae: This is the celestial equivalent of a Latitude on Earth, which when projected to the ***CS*** defines the corresponding ***Decl***. See figure **8 Celestial Coordinates**.

Equation of Time: This is the difference between ***Apparent Solar Time (ASolT)*** and ***Mean Solar Time (MSolT)***, that is:

> ***ASolT – MSolT = Equation of Time*** obtainable from an Internet calculator e.g.
> https://keisan.casio.com/exec/system/1271898403

> You input the Longitude e.g. -81, the time zone e.g. -5 and out comes a list of ***Equations of Time*** for the whole year at intervals of X days which you specify e.g. 5.

Great Circle: A great circle is an important component of the measurement grids of the sky. It is a circle drawn on a sphere, having the same centre as the sphere's e.g. the Equator is a great circle on Earth's sphere, as is its equivalent the ***CE*** on the ***CS***. Another example is a great circle passing through the ***CNP***. By definition, it also passes through the ***CSP***, and represents an ***RA***, 'an' because one can draw numerous great circles at the ***CNP***. But note must be made about the other important circles, Latitudes and ***Decls***. These are NOT great circles because their centres are not coincident with a sphere's. That's why they are called the *parallels* of Latitude, meaning that starting at the Equator (which IS a great circle) these are circles parallel to it, meaning their centres are up and down the Earth's rotational axis. An important great circle is the **Zenith** great circle. As the name implies, it is a great circle passing through an observer's **Zenith**.

Greenwich Mean Time (GMT): This is the time zone in Great Britain. It follows **UTC** but also adjusts one hour for daylight saving, written as **GMT**+1, which **UTC** doesn't.

Hour Angle: In very general terms, this is an angle between the planes of any two meridians (or great circles) of the Earth or the **Celestial Sphere** and is measured along the sphere's Equator in the *westerly* direction. When these meridians refer to specific ones such as the Greenwich meridian and a celestial meridian through a star, it then has a specific name which for this example is the Greenwich Hour Angle. Another example is **Local Hour Angle**.

Local Hour Angle (LHA): This is an hour angle between the **LM** and a star's celestial meridian.

Local Meridian (LM): This is a local longitudinal meridian of an observer projected on to the **CS**. Whereas the local Longitude is fixed on the ground, its projected **LM** moves across the **CS** as the Earth rotates.

Magnitude: With respect to sky objects, it refers to their apparent brightness as observed from Earth and is expressed as a number on a logarithmic scale, with star Vega at the origin of this scale i.e. having a value of zero. Objects dimmer than Vega are assigned positive values, that is the higher the positive value the dimmer the object. Objects brighter than Vega are assigned negative values, the higher the absolute value of negative values the brighter the star. There are only *four* stars (plus the Sun and the Moon) brighter than Vega (see list below), and of the planets, only the largest, Jupiter, and the two between Earth and the Sun, Venus and Mercury, are brighter than Vega. Some magnitudes to compare: [stars are flagged *]

-26.8 Sun
-12.7 average Full Moon
-4.7 Venus
-1.8 Jupiter
-1.5 *Sirius
-0.7 *Canopus
-0.4 Mercury
-0.3 *Rigel Kentaurus
-0.05 *Arcturus
0 *__Vega__
0.1 *Capella
0.2 *Rigel
0.34 Mars
0.49 Saturn
< 1.5 **1st order of _Magnitude_** limit (the 21 brightest stars)
>= 1.5, < 2.5 **2nd order of _Magnitude_** limit, and so on
1.95 *Polaris
5.75 Uranus
6-7 **naked eye limit**
7.82 Neptune

**Mean Solar Time (MSolT)**: This is the normal time in everyday use, as compared to sundial time which uses the apparent sun to show the 'true'/apparent local time. The former uses a clock which simulates a mean sun of constant cycles. The apparent sun does not move at a constant rate, but an accurate clock does and represents _**MSolT**_. Each time zone has its own _**MSolT**_ which is the same throughout the whole zone, regardless of local Longitudes. _**MSolT**_ is assigned as its zone meridian/Longitude divided by 15° plus the current _**GMT**_ e.g. for the Eastern time zone in North America, its _**MSolT**_ is: -75 ÷ 15 + _**GMT**_ = _**GMT**_ − 5.

Planetary terminology:
   **Conjunction** is when a planet and any other celestial object appear to be meeting.

Elongation is when a planet is at the maximum apparent distance from the Sun, as observed from Earth. See figures **33 Planets in View Nov. 6, 2018** and **37 Planets in View Aug. 6, 2018**.

Occultation occurs when a planet eclipses another celestial object.

Opposition is the state where two celestial objects are on the opposite sides of the **Celestial Sphere**. Since the orbits of **Inferior Planets** are entirely in front of Earth, they are always on the same side of the **CS**, and therefore never achieve **Opposition**.

Inferior & Superior Planets: The former are closer to the Sun than Earth (viz. Mercury and Venus), the latter further away (viz. Mars, Jupiter, Saturn, Uranus and Neptune).

Inner & Outer Planets: the former are the ones between the Asteroid Belt and the Sun (viz. Mars, Earth, Venus and Mercury), the latter outside the Asteroid Belt (viz. Jupiter, Saturn, Uranus and Neptune).

Right Ascension (RA), designated as α (alpha) in formulae: This is an **Hour Angle** between the planes of the celestial meridians of a star and of the **Vernal Equinox**, measured in the *easterly* direction.

Plan View of CS

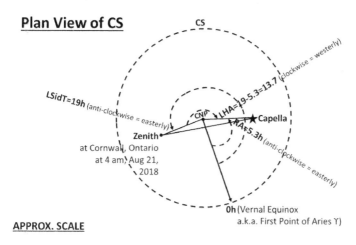

69 Different CNP Angles seen from Plan View of CS

The above figure shows a plan view of the **CS** with the **CNP** in the centre. It allows us to understand different angular measurements and the directions in which they are measured e.g. **RA** is measured anti-clockwise (easterly direction) from the Vernal Equinox to the object of interest (star Capella in this case). Similarly, **LSidT** is also measured anti-clockwise (easterly direction) from the Vernal Equinox to the object of interest (the observation **Zenith** in this case). While the **RA** remains the same, **LSidT** changes continuously as the Earth rotates. As opposed to the above two angles, an **LHA** is measured clockwise (westerly direction) from the **LM** to the object of interest (star Capella).

Sidereal Time: This is star-based time, as opposed to sun-based (solar) time. A sidereal day is the time it takes for a star to apparently revolve around the sky and return to the same position the next day, a sidereal day being about 23h 56m long, as measured using our normal 24-hour clock. The average position of the **Vernal** (Spring / March) **Equinox** is the zero point of **RA**s, meaning its **RA** is 0h, thus the **Sidereal Time** is 0h too (the **RA** of transiting body is equal to the **Sidereal Time** at that instance). About every 23h 56m of **MSolT**, sidereal time goes back to zero and another sidereal day begins anew. Thus, the sidereal day goes ahead, every solar day, by about 4m, and at the end of the year, it will be one whole day ahead. How this happens is seen in the diagram below: [Recall, the Earth rotates in the *easterly* direction, which in this view below, it is rotating anti-clockwise, so on day 2, it will see the distant star *before* seeing the closer sun]

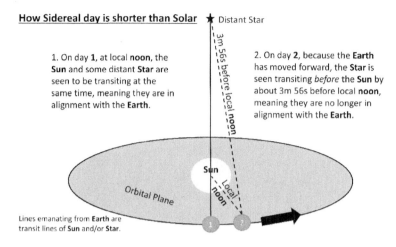

How Sidereal day is shorter than Solar ★ Distant Star

1. On day **1**, at local **noon**, the **Sun** and some distant **Star** are seen to be transiting at the same time, meaning they are in alignment with the **Earth**.

2. On day **2**, because the **Earth** has moved forward, the **Star** is seen transiting *before* the **Sun** by about 3m 56s before local **noon**, meaning they are no longer in alignment with the **Earth**.

3m 56s before local noon

Sun

Local noon

Orbital Plane

1 2

Lines emanating from **Earth** are transit lines of Sun and/or **Star**.

70 Sidereal Day

Standard Time: This is a time zone's **MSolT** and is equal to **UTC±n** e.g. Eastern Standard Time (EST) = **UTC**-5. When a standard time is adjusted for daylight saving, it is written as Daylight-saving Time e.g. EDT = **UTC**-4.

UTC: Universal Time Coordinated, also known as Coordinated Universal Time. This is the **MSolT** at Longitude 0°. It is equal to **GMT**, except during daylight saving when an hour is added to **GMT**.

Vernal Equinox (VE): This is a point on the **CE** where the apparent sun crosses it around March 20, crossing it from the southeast to the northwest. It is the 0h reference point for **RA**s. [Greek astronomy called this the First Point of Aries ♈ since at the time this crossing point's background constellation was Aries the Ram ♈. However, because of Earth's Precession, the **VE** does not occur at the same point on the **CE**, and currently its background constellation is Pisces.] Being on the **CE**, the celestial coordinates of the **VE** are **RA0h, Decl0°**, which makes it the origin point of the Celestial Equatorial Coordinate System.

Zenith (Z): This is a point in the sky directly above an observer. It sits on the local Latitude projected on to the **CS** where it is given the name ***Declination***. In other words, the local Latitude is equal to **Z**'s ***Declination***.

List of Key Facts

CNP is at the **handle's tip** in the Little Dipper/ladle of Ursa Minor. This is where **Polaris** is located. It is visible from almost anywhere in the northern hemisphere.

The far edge of the Big Dipper of Ursa Major points to **Polaris**. This edge is made up of two stars, Merak and Dubhe, known as **pointer stars**.

The long arm of the Crux constellation, roughly pointing to the *CSP*, is made up of two stars, Gacrux and Acrux, known as **pointer stars**. The *CSP* is visible from almost anywhere in the southern hemisphere.

The Big Dipper is 'visible' from anywhere in the northern hemisphere, ditto the Crux in the southern ('visible' in quotes to indicate that although these are up there in the sky, they are truly visible at night only). From about Latitude 40°N to the *CNP*, the Big Dipper is *Circumpolar*, and from about Latitude 35°S to the *CSP*, the Crux is *Circumpolar*, meaning these asterisms are up in the sky all night long, never rising nor setting, thus no waiting is required for them to show up (rise). But, between each of the above Latitudes and the Equator, one has to wait for the asterisms to rise :-(To find out the viewing time window for one of these asterisms, use a tool like Stellarium, making it spin the sky through the night and noting the rising and setting of the asterism – see **Appendix F: Night Sky Tool**.

By definition, LSidT equals zero where the Sun is transiting at *VE*, since its *RA* is zero then.

The northerly end of **Orion's belt** almost touches the *CE*. Other bright stars close to the *CE* are:

Sirius, the brightest star, is about 17° South of the **CE**.
Rigel, in Orion, about 8° South of the **CE**.
Procyon is about 5° North of the **CE**.
Altair is about 9° North of the **CE**.
Spica is about 11° South of the **CE**.
Aldeberan is about 5° North of the **CE**.

Use these stars to locate the **CE** in the night sky.

The arc-distance from the **CE** to the **Zenith** is the **local Latitude = Declination**. Conversely, if the local Latitude is known, then its arc-distance can be used to locate the **CE**.

The **LM** is equal to LSidT which also is its **RA** at that exact point in time, which then is a good initial reference. Although it is moving, it can be updated by the elapsed time, with good approximation.

The **LM**'s **RA** equals the **RA** of a star in **transit**.

Earth rotates from **West to East**, so stars appear to rotate the other way. In other words, they rise in the East and set in the West (as do the Sun, the Moon and the planets).

RA is measured on the **CE** in an **easterly** direction from the **VE** (same as for LSidT). **LHA** on the other hand is measured in a **westerly** direction from the **LM**. See diagram **69 Different CNP Angles seen from Plan View** of CS, where it is seen that **RA**s are measured in an **anti-clockwise** direction.

RAs are like clock times, except they follow **Sidereal Time**. Thus, if the current LSidT is say 12h, then say star Antares, having **RA16.5**, will transit 4 ½ hours later. [Wouldn't it be nice if there were watches that told **Sidereal Time**? Well, that's exactly what a smart phone app does, like mine.]

RAs are **stationary** on the **CE**, while Longitudes are **stationary** on Earth. An observer on Earth sees the **RA**s moving while the local Longitude remains stationary. Likewise, an ET (extra-terrestrial) on the **CE** would see Earth's Longitudes moving while its local **RA** remains stationary.

Except for the Equator, Latitudes are not **great circles**, as all Longitudes are.

At any time, our view of the sky is a half-shell (**dome**) of the **CS**, with the **Zenith** being at the shell/dome's apex. The northern **horizon** is L° degrees off the **CNP** (whose **Decl** is +90°) and the southern, L° off the **CSP** (whose **Decl** is -90°) where L is the Latitude of observation e.g. if L is +30°, then the Northern horizon will be **CNP** +30°, the southern, **CSP** +30°; and if L is -30°, then the Northern horizon will be **CNP** -30°, the Southern, **CSP** -30°.

The **CNP** or the **CSP** should be at an **Altitude** equal to the absolute value of the local **Latitude**.

The **CE** meets the horizon at its **East** and **West cardinal** points.

Looking up at the **CNP** from inside the **CS**, stars rotate around it in an **anti-clockwise** direction. Looking at the **CSP** from inside the **CS**, stars rotate around it in a **clockwise** direction. But, in both cases they are moving in a *westerly* direction.

Stars around the **Zenith** look **brighter than their rankings**, and above the horizon, dimmer than their rankings, because of atmospheric attenuation.

GMSidT – LMSidT = local **Longitude**.

Celestial coordinates of the **VE** are **RA0h, Decl0°**.

Stars rise in the East **four minutes** earlier every night. Thus, everyday **Sidereal Time** gains on regular clocks by about four minutes, which means it gains about a **couple of hours every month**, very useful in estimating when a star or constellation will be locally in view e.g. if you are observing at 8 pm and LSidT is zero, it means Orion (whose **RA** is just under 6h) is currently at the horizon and so is not good for observation, but in a month's time it will be above horizon by a couple of hours i.e. its **Altitude** will 30° (2h X 15).

A **blackout** zone is around the celestial pole furthest from you and has a radius equal to the absolute value of your Latitude.

RA is like a 'timestamp', meaning it is fixed (permanent) just like a timestamp. As compared to this, your 'wristwatch' (i.e. your smartphone *Sidereal Time* app) is ticking. Both, the timestamp and the wristwatch, keep to *Sidereal Time*, with the difference that the timestamp is permanent while the wristwatch is ticking (your LSidT is always ticking). E.g. if a timestamp is *RA19.8h* (which is star Altair's *RA*) and your wristwatch says the LSidT is 15.0h, then Altair will arrive / transit in 4.8h (19.8 – 15.0).

Any of the stars close to the *CE* also serve as *RA* markers e.g. star Mintaka, the top of Orion's belt, just touches the *CE* at *RA5.5h*. That is, Mintaka serves as the marker *RA5.5h* on the *CE*, from which other markers can be laid out to assist in finding other objects.

Appendix A: Obtaining Local Sidereal Time

Your best bet in obtaining Local *Sidereal Time* (LSidT) is a smart phone app. I have used one to my satisfaction. **Tip**: My app is like a ticking watch, the times it displays (LSidT, *Standard Time* and *UTC*, by default) are in ticking mode which you can only pause. So, how does one find the *Sidereal Time* for a given *Standard Time?* By pausing the app, noting the difference between the displayed *Standard Time* and *Sidereal Time*, and applying that difference to the desired *Standard Time*. For example, my app shows the following, right now:

Standard Time 10:51:47

Sidereal Time 18:50:00 [i.e. 7:58:13 more than the above]

Then, say you want to know the LSidT for 9 pm tonight. You add the above difference to 9 pm (21:00:00), getting 4:58:13 (after subtracting 24 hours). This is approximate, because every 24 hours, *Sidereal Time* is ahead of *Standard Time* by about 3m 56s ($\approx$ 4m), meaning about 4m must be added on top of the above difference, for every 24 hours. For our purposes, within a 24-hour period, this correction can be ignored.

However, if you need an Internet calculator to do this, here are a few examples: [in all of these, a form is filled out and a *Sidereal Time* is calculated, among other things]

https://eco.mtk.nao.ac.jp/cgi-bin/koyomi/cande/gst_en.cgi [calculates GSidT, also *Equation of Time*]

http://jukaukor.mbnet.fi/star_altitude.html [which is the same calculator as in **Appendix B: Converting Celestial Coordinates to Local Horizon Coordinates**, and therefore requires more input as compared to the first calculator which only requires date and time, however this second calculator produces both GSidT AND LSidT.]

Lastly, if you don't have access to any of the above means of calculating, you can obtain it approximately (within a couple of minutes or so) by interpolation of the following approximate values, but it is not so simple as is seen in the example below:

Approximate GsidT at 0 am on 1st of each month

January	6.69h
February	8.7h
March	10.6h
April	12.6h
May	14.6h
June	16.6h
July	18.6h
August	20.6h
September	22.7h
October	0.6h
November	2.7h
December	4.6h

Example:

June 21st, 2015 4:30 am, Latitude 44.5°, Longitude -80.2°, Collingwood, ON.

We need to interpolate the above values for June 1st and July 1st:

GsidT diff. June 1st & July 1st:	18.6 – 16.6	= 2
Convert Local Time to UTC:	4:30 + 5 – 1*	= 8:30
	*DST adjustment	
Time diff. observation & 0 am:	8:30 – 0	= 8:30 = 8.5
Sidereal correction:	8.5 * (1 + (3.93 ÷ 60) ÷ 24)	= 8.5231

Appendix A: Obtaining Local Sidereal Time XE "tip:how to study"

#days from June 1ˢᵗ to 21ˢᵗ:	21 – 1	= 20
#days from June 1ˢᵗ & July 1ˢᵗ:	30 – 1 + 1	= 30
Interpolation ratio:	20 ÷ 30	= 0.6667
Interpolated value:	16.6 + (0.6667 X 2)	= 17.9334
Value at UTC 8:30:	17.9334 + 8.5231	= 26.4565
Convert to LSidT:	26. 4565 + (-80.2 ÷ 15)	= **21.2**

Looking up this value in **Table of Coordinates: by RA-Decl** in **Appendix L: Information Tables**, we see that the nearest bright star is Deneb having **RA20.7**, that is a difference of (21.2 – 20.7) = 0.5h, meaning Deneb already transited 0.5h ago. The difference between Deneb's **Decl45°** and the Latitude 44.5° is 0.8°, that is Deneb is that many degrees higher in **Altitude** than the **Zenith**. All this is precisely captured in the observation photo **65 Collingwood Observation Photo**. The photo also shows Vega (**RA18.6**) transited (21.2 – 18.6) = 2.6h ago, and that Alderamin (**RA21.3**) will transit in (21.2 – 21.3) = -0.1h, and so on.

Appendix A: Obtaining Local Sidereal Time XE "tip:how to study"

Appendix B: Converting Celestial Coordinates to Local Horizon Coordinates

Celestial coordinates (*RAs* and *Decls*) are used for star positions on a spherical grid anchored by Earth's Equator projected on to the *CS* and by a point on the *CE* where the Sun crosses it during a *VE* e.g. star Vega's celestial coordinates are *RA18.6h* and *Decl39°*. These coordinates are sky-based and thus independent of an observer on ground, and whose view for a given date-time keeps changing, as the Earth rotates and orbits around the Sun. Thus, for this observer, the celestial coordinate system has to be converted to a local horizon-based system as seen in figure **9 Local Horizon Coordinates.**

There are calculators on the internet to do this conversion. One such is http://jukaukor.mbnet.fi/star_altitude.html.

Like many calculators you fill in a form, specifying a celestial object's coordinates, your location (Latitude, Longitude) and the observation date-time (given as *UTC*), and out come your Local Horizon Coordinates (*Azimuth* and *Altitude*) plus additional info such as *Sidereal Time*.

Appendix C: Converting Local Horizon Coordinates to Celestial Coordinates

This is the reverse of **Appendix B: Converting Celestial Coordinates to Local Horizon Coordinates** e.g. you observed something interesting in the sky and measured its Local Horizon Coordinates (*Azimuth* and *Altitude*), and now you want to know what its celestial coordinates (*RAs* and *Decls*) are. Again, a calculator on the internet can be used e.g.

https://clearskytonight.com/projects/astronomycalculator/coordi nate/horizon_equatorial.html

Here too, the *Azimuth* and *Altitude* are input, plus the Latitude of observation and out come – the *Declination* BUT not the *RA*, instead of which there is a (*Local*) *Hour Angle*. To convert *LHA* to *RA*, the LSidT (Local Sidereal Time) of observation (obtained say from a smart phone app) is required. Thus,

RA = LSidT – *LHA* [if result is negative, add 24 hours]

Appendix D: Celestial & Galactic Coordinates of Stars

The Simbad database at the University of Strasbourg, France, is one of the good sources for Celestial and Galactic Coordinates:

http://simbad.u-strasbg.fr/simbad/sim-fid

Basically, you type in an identifier (e.g. Altair) and out come the coordinates, among other goodies.

If you need to convert between Celestial and Galactic Coordinate Systems, use a suitable calculator on the Internet e.g.

https://ned.ipac.caltech.edu/coordinate_calculator

Appendix E: Planet RA & Declination

In addition to using a graphic tool such as Stellarium (see **Appendix F: Night Sky Tool**) to look for when a planet(s) that you want to observe will be in the night sky (e.g. by making Stellarium's display go fast-forward), you can use calculators on the Internet to obtain viewing times.

https://www.timeanddate.com/astronomy/night/

"Time and Date" automatically shows you the viewing times for today for your location. You can change location and/or date-time to look for your planet(s).

If you want coordinates of the planet(s), again Stellarium will do the job, but if you must use a tool on the Internet (e.g. you have not installed Stellarium), here's one:

https://keisan.casio.com/exec/system/1224748262

Appendix F: Night Sky Tool

Here we take a look at a couple of rendering tools. For our purposes they are most useful in preparing for observation or post-observation identification. And, if for any reason you do not want to make observations (e.g. too much light pollution), these tools can be substitutes for it (a.k.a. armchair astronomy).

We will look at the tools in order of their age:

Fourmilab's Your Sky: http://www.fourmilab.ch/cgi-bin/Yoursky

This was developed by John Walker, founder of the computer company AutoDesk, makers of the well-known AutoCAD software. You type in your location and date-time of observation and it will render your sky, showing celestial objects. You can also type in what and how much you want to see. For example, the following criteria produces the screen shot below:

Universal Time: **2019-01-06 11:00:00**
Latitude: **25° North**, Longitude: **80° West**
Ecliptic and Equator box **checked**
Moon and planets box **checked**
Show stars brighter than magnitude: **2.5**
Names for magnitude brighter than: **2.0**

In this image, we see stars that we have become familiar with in previous chapters. It also illustrates why the need for what and how much to see: in this image, there isn't any clutter except near the *CSP*, which can be adjusted as need be.

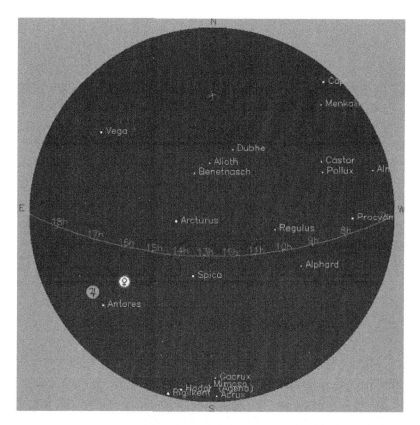

71 Screen shot for above parameters (from "Your Sky" by Fourmilab)

Stellarium: https://stellarium.org/

This is a freeware desktop application developed by the stellarium.org group. It comes with a user manual. It has to be downloaded and installed on your computer. When it is run, it automatically determines your location, sets the date-time to now and displays the current sky. If it is day time, it renders it quite realistically (not much to see though, as would be in reality, except for the Sun, the Moon and perhaps planet Venus), but you can change it by clicking on the "Atmosphere" toggle icon to turn off atmospheric scattering (which scatters sun's blue / violet light, creating a blue sky and making stars invisible). There's a ton of stuff you can do with Stellarium, all quite intuitive. A very good example is simulating the passage of time by clicking the "Increase / Decrease time speed" icons. This makes the sky rotate at your desired speed. Very cool. Click on a celestial object and it gives you lots of info including object name, *RA-Decl*, *Azimuth-Altitude*, *Sidereal Time*, etc. In addition to its GUI (graphical user interface), it has keyboard shortcuts for better productivity. Here is a view from it, equivalent to the one above:

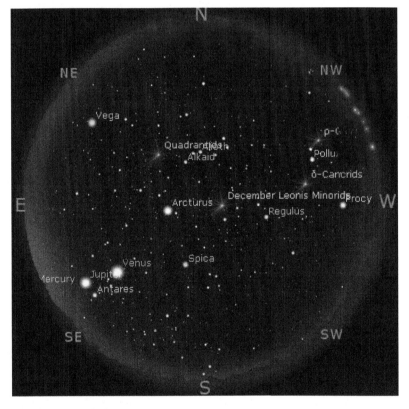

72 Screen shot (from "Stellarium")

Note: The clutter near the South horizon in the previous screenshot is not seen above, as Stellarium allows for atmospheric attenuation and accordingly downgrades the stated **Magnitude**. Attenuation is maximum near the horizon, hence the stars near the South horizon are not displayed because of **Magnitude** cut-off.

In addition to the objects seen in "Your Sky", it shows meteor showers e.g. "December Leonis Minorids" (between Arcturus and Regulus in the above image). This product is chockful of goodies and everything that is displayed is adjustable, including turning on grid lines (celestial coordinate system and / or local horizon system).

For how to use the most common functions of Stellarium, see **A Complete Observation Package** in Appendix K: Preparing for Observation. [Among the functions described there, is rotating the horizon to match what one was, or will be, facing in the field, most useful in customizing a view cf. Fourmilab's "Your Sky" let's you specify the horizon direction as only North or South. This is a good illustration of the difference in the state-of-the-art before and after the turn of the millennium.]

Most of our needs are satisfied by tools like the above, but sometimes there are additional features some others provide. A good example is "In the Sky": https://in-the-sky.org/skymap.php. Among other things, it tells you about all those satellites up there in orbit and which are visible to the naked eye, looking like stars on the move! It tells us when we can see them at any given location, graphically showing the paths in sky. This is most necessary because they move fast across the sky, taking only minutes to do so (and visible only when they are not in Earth's shadow), thus it tells you the from-to time window of view so you can be at the ready. Another nice thing it provides is a digital orrery. As the saying goes, "A picture is worth 1,000 words", so it is the case here: you can see planets in motion, month after month, change your perspective, zoom in / out, etc. It is Planetary Motion 101 needing no words!

Appendix G: Cosine Rule of Spherical Triangles: Arc-distance Calculation

A spherical triangle is between three points on the surface of a sphere such as the celestial sphere e.g. the Summer Triangle of Vega-Altair-Deneb. This being a curved surface, the triangle sides are arcs of the *Great Circles* joining the three points. Thus, the lengths of the triangle's sides are expressed as arc-angles at the centre of the sphere. The cosine rule applies trigonometric functions to these angles.

How is the Cosine Rule applicable to us naked eye astronomers? Frankly, it is not, because its application is in calculating various things and we have seen that naked eye astronomy is doable without any complicated calculations. So why this appendix? It's for the curious or interested among us, wanting to delve into some mathematics behind astronomy. The Cosine Rule is a good introduction.

We'll use an example similar to the one in figure **69 Different CNP Angles seen from Plan View of CS**. The figure below is an example of the Cosine Rule for calculating the arc-distance between two stars:

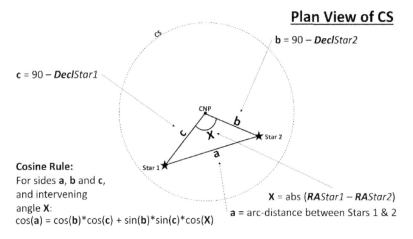

Plan View of CS

$b = 90 - DeclStar2$

$c = 90 - DeclStar1$

CNP

Star 2

Star 1

Cosine Rule:
For sides **a**, **b** and **c**,
and intervening
angle **X**:
$\cos(a) = \cos(b)*\cos(c) + \sin(b)*\sin(c)*\cos(X)$

$X = abs\ (RAStar1 - RAStar2)$
a = arc-distance between Stars 1 & 2

73 Cosine Rule

Here's an example using the above formula:

Star 1 = Arcturus *RA14.3h, Decl19°*

Star 2 = Vega *RA18.6h, Decl39°*

$\cos(a) = \cos(90-19)*\cos(90-39) +$

$\sin(90-19)*\sin(90-39)*\cos(15*abs(18.6-14.3))$

$= 0.521$

Therefore, **a** = acos(0.523) = **59°** = arc-distance between Arcturus
and Vega!

There are calculators on the Internet to calculate arc-distances
between two sets of *RAs, Decls* e.g.

http://www.gyes.eu/calculator/calculator_page1.htm

Appendix H: Determining Local Latitude & Longitude

This is not a challenge any more (as Longitude was a few centuries ago) given satellites and accurate clocks that let you triangulate your position and Internet technology making it all readily available. But, it is satisfying to be able to measure these by ourselves, albeit to an approximate level:

Latitude: Locate the **CNP** or the **CSP**. Measure the arc-distance from it to the **Zenith**. Subtract it from it 90. Voila! Measuring the pole star's **Altitude** is another approach and which equals your Latitude. Alternatively, a sextant can be used – see **Appendix I: Home-made Sextant** – to measure the angle without reference to horizon or **Zenith**.

Longitude: This is not as easy as the above. You have to find a star (East of the **Local Meridian**) that you can recognize from previous experience, which is very doable given enough experience. [Though the **CNP** or the **CSP** are easy to locate they can't be used as they don't transit – they are stationary.] Say you recognize constellation Orion whose brightest star Rigel, at the hunter's front knee, happens to be East of the **LM**. Now, you have to wait for it to transit. When it does*, the LSidT at that time is equal to the Rigel's **RA**, being 5:15:28. For example: [*see **Appendix J: Measuring Transit**]

On 2019-01-27, Rigel's Transit Time was 21:07:51.

At 9:40 am local time, GMSidT was recorded as 23:06:21 (see **Appendix A: Obtaining Local Sidereal Time**).

Thus, GMSidT at local transit time 21:07:51 is equal to:

(21:07:51 − 9:40:00) + 23:06:21 = 11:27:51 + 23:06:21 = 34:36:05

Plus + 0:01:53* = 10:36:05 [*sidereal correction: 11:27:51/24 times 3m56s]

Therefore:

Local Longitude = LSidT − GSidT = 5:15:28 − 10:36:05 = -5:20:37 = 80.15°West (West is negative).

The above required obtaining GMSidT from the Internet. But what if you are in a dale somewhere in Shangri-La where you may not have wireless Internet? If you have a satellite phone, you could call your buddy who could tell you the current GMSidT. Or, to save the cost of a satellite phone call, you can record the GMSidT before leaving, similar to the above example, and not forgetting to apply a sidereal time correction (as seen above) for time elapsed between when the GsidT was recorded and the present.

Using the Sun to Determine Longitude: This approach determines the local apparent solar noon (see figure below), then calculates the difference between it and the local time zone's mean solar noon *MSolT*.

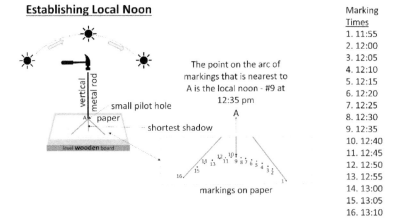

Establishing Local Noon

The point on the arc of markings that is nearest to A is the local noon - #9 at 12:35 pm

vertical · metal rod

small pilot hole

paper

shortest shadow

level wooden board

markings on paper

Marking Times
1. 11:55
2. 12:00
3. 12:05
4. 12:10
5. 12:15
6. 12:20
7. 12:25
8. 12:30
9. 12:35
10. 12:40
11. 12:45
12. 12:50
13. 12:55
14. 13:00
15. 13:05
16. 13:10

74 Local Apparent Noon

Appendix H: Determining Local Latitude & Longitude

In the above figure, apparent solar noon has been measured as 12:35 pm (**ASolT**). The local time zone's solar noon was at 12 o'clock (**MSolT**), meaning our local noon is behind clock noon by 35 minutes. If this was EST (Eastern Standard Time), the zone's meridian is 75° West of Greenwich, where at noon the Sun is at its highest altitude. Converting the 35 minutes to degrees:

35 minutes = 0.583 hours = 8.75 degrees (hours X 15).

Thus, local Longitude = 75 + 8.75 = 83.75°, less *****Equation of Time** correction of -13m25s (-3.35°) = 80.4° West of Greenwich, which is pretty close to the accurate Longitude of 80.15°. [*needed to obtain the exact time of noon at the zone meridian since it fluctuates from the steady **MSolT**.]

Appendix I: Home-made Sextant

A sextant can be used to measure the **Zenith** angle without reference to Horizon or **Zenith**.

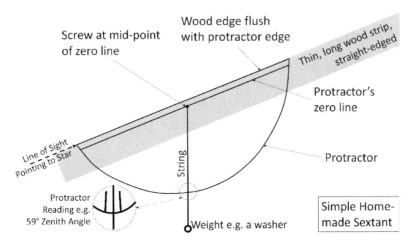

75 Home-made Sextant

The above home-made sextant is used to point to an object in the sky along its line of sight. The plumb line (string) is pinched with thumb and fore-finger to hold it in place for reading the protractor's degree marking, which produces a **Zenith** angle which when subtracted from 90 produces **Altitude**.

Trivia: A modern constellation is named after this simple but very useful instrument.

Appendix J: Measuring Transit

Measuring Transit: This is easier said than done, because the *Local Meridian* across which the transit takes place is only an imaginary line in the sky. One way is to construct a part of the *Local Meridian* on the ground as seen in the figure below:

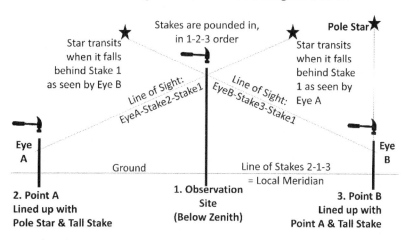

76 Measuring Transit

When a star whose transit you want to observe disappears behind the middle observation stake (as explained above), you record the time then, as well as the time when it emerges back into view on the other side of the stake. The average of these two times is the transit time. At that time, the LSidT is equal to the star's **RA**.

Appendix K: Preparing for Observation

The preparations discussed below are a range from no preparations to progressively more. **Tip**: Keep a small notebook, in which to make pre- and post-observation notes and sketches. Number the pages for use in a table of contents and for cross-referencing. All observation notes and sketched should have date-time and location.

No Preparations

This is where you simply want to step out and observe what the sky holds for you and make notes. For example, one sees one of the brighter stars (Sirius, Canopus, Rigel Kentaurus, ...) and wonders what it is:

At **Lat26°, Long-80°**, at 5 am Jan 24, 2019

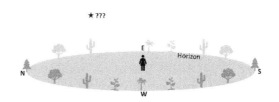

APPROX. SCALE

77 Unknown Star Observed

That's an observation, for which you want to make notes for later identification of the star:

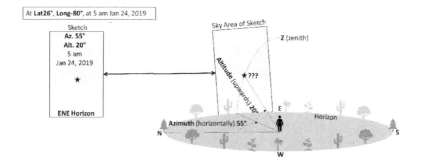

78 Sketch with Azimuth & Altitude

The **Azimuth** and **Altitude** (called Local Horizon Coordinates) measured above are converted to celestial coordinates as shown in **Appendix C: Converting Local Horizon Coordinates to Celestial Coordinates**, which produces the following:

Local Hour Angle*: 18.4h

Declination*: 39°

From which we get (again, as shown in the above appendix):

RA*: LSidT – **Local Hour Angle** = 12.9 – 18.4 = -5.5 + 24 = 18.5h

LSidT is obtained as shown in **Appendix A: Obtaining Local Sidereal Time**.

Finally, we use these coordinates to find a close match with a row in **Table of Coordinates: by RA-Decl** in **Appendix L: Information Tables**. The closest match is:

Vega: **RA18.6h, Declination39°**, which also is the 5th brightest star.

What if there is no close match? This can happen if it is not among the bright stars in the above look-up table, and yet stands out if it is sufficiently high up in altitude (meaning, atmospheric attenuation is relatively low). For example, after converting the measured Local Horizon Coordinates of such a star to celestial, you come up with **RA17.5h, Declination13°**, but assume no close match is found in the above look-up table. Then the quest is to visually match it in a rendered sky using a suitable tool. For example, we bring up Stellarium, setting it to the date-time and location of observation, and rotating the horizon to the point above which the star was observed, then turning on the display of the celestial coordinates grid lines. You find a star close to the measured celestial coordinates. Click it and voila – the star is **Rasalhague** (which happens to be the 57ᵗʰ brightest and actually is in the look-up table):

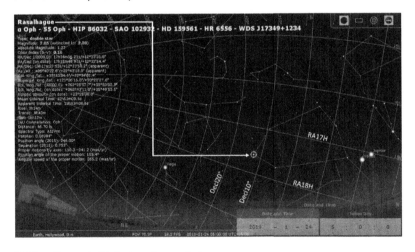

79 Rasalhague in Stellarium Screenshot

Preparations for Finding Orion

We start with Orion, being visible from almost all countries on Earth since it is on the **CE**. Also, we have already taken a look at it

– see figure **28 Orion**. To locate it in our sky, we use both approaches previously seen: **(1) Co-ordinate Approach** and **(2) Graphic Approach**.

The challenge here is that Orion is in the night sky for only part of a year, and in the day sky for the remainder, for any given location. We will use the coordinate approach first. The top of Orion's belt is star Mintaka which almost touches the **CE** as per its **Decl-0°**. But it is its **RA5.5h** that we will use for finding a suitable night time view, and the fact that when it is in transit, its **RA** is equal to LSidT. Say the current LSidT (which is obtained as shown in **Appendix A: Obtaining Local Sidereal Time**) is 22:26:30 (≈ 22.5h) and current local time is 14:27:30 (≈ 14.5h). The LSidT tells us that Mintaka's transit is about 7 (5.5 − 22.5 + 24) hours away (ignoring the sidereal correction of 4m every 24h). Thus, the local time of transit will be about 9:30 pm (14.5 + 7). At that time, Mintaka will be crossing the **Local Meridian**, and at the **CE** to boot, making it easy to locate it, as previously seen in **Trace the CE** in **10 Walkabouts in the Night Sky**. The Transit Time tells us that Mintaka rose at 3:30 pm (9:30 pm − 6) and will set at 3:30 am (9:30 pm + 6). If sunset is at say 6:30 pm (which is later than Mintaka's rise at 3:30 pm), Orion will already be up high enough for us to easily identify it and it will still be up enough past midnight.

But, months from now, Orion will be in day time sky. For example, six months from now, the above calculation would reveal that Mintaka will transit during the noon hour. So then, we need to work out when it will be in the night sky, as compared to the above example in which it was but by sheer coincidence.

The way to do this is to start with the LSidT for say the first of the month at 9 pm e.g.

January 1, 2019 at 9 pm: LSidT = 3.4h. Longitude -80°.

Thus, Mintaka's transit is:

RA5.5h – 3.4 = 2.1h after 9 pm = 11.1 pm, rising at 11.1 pm – 6h and setting at 11.1 + 6h, the rising and setting being the maximum window of time that Mintaka will be above the horizon. Thereafter, it will transit roughly 2h earlier every month. These are very rough figures but good enough for planning purposes. A monthly table can be built as follows: [day time transit greyed out]

Month	Transit	Rise	Set
2019-1-1	11.1 pm	5.1 pm	5.1 am
2019-2-1	9.1 pm	3.1 pm	3.1 am
2019-3-1	7.1 pm	1.1 pm	1.1 am
2019-4-1	5.1 pm	11.1 am	11.1 pm
2019-5-1	3.1 pm	9.1 am	9.1 pm
2019-6-1	1.1 pm	7.1 am	7.1 pm
2019-7-1	11.1 am	5.1 am	5.1 pm
2019-8-1	9.1 am	3.1 am	3.1 pm
2019-9-1	7.1 am	1.1 am	1.1 pm
2019-10-1	5.1 am	11.1 pm	11.1 am
2019-11-1	3.1 am	9.1 pm	9.1 am
2019-11-1	1.1 am	7.1 pm	7.1 am

Based on your preferred observation time, you can pick a suitable month, reasonably close to transit time so that Mintaka is high in the sky (meaning it is least affected by atmospheric attenuation). For example, if you like to observe at 9 pm, around the beginning of February is a good time, although the months of January and February are okay enough. Once you have picked dates (plus rain / cloud dates), you can calculate the precise transit time for those dates, and you're all set to go.

The above was the coordinate approach. The **graphic approach** uses a rendering tool to inspect the current sky, hour-by-hour or day-by-day or month-by-month. For example, Stellarium (**Appendix F: Night Sky Tool**) automatically opens with a sky view of the current location and date-time, showing a southerly portion of the local sky, which can easily be changed to a full view by zooming out. You can easily change the location by selecting a city anywhere in the world: say you are going to Arusha, Tanzania for a tour of its famous game parks, including the Ngorongoro Crater (the Garden of Eden) and the Olduvai Gorge where humankind was born, hoping to squeeze in some night observations in equatorial East Africa. It will be at the top of Stellarium's list after you have typed a few letters e.g. "Arus". Click it and the location changes to its Latitude and Longitude. But herein we will stick to our current location for now. If your intended viewing time is different, you can easily change it with mouse or keyboard. Let's say you want to observe the stars of constellation Orion but it is not currently in view in Stellarium. All you need to do is increase or decrease the current time (by hours). If you are a night owl, you wouldn't mind staying up late, so you can increase the time until Orion comes into view. Hopefully, in a short time, its two brightest stars Rigel and Betelgeuse will appear, with its iconic 3-star belt clearly seen between them. But, it could also happen that Orion is in the daytime sky, currently, in which case you would need to increase or decrease dates and look for a suitable period when it is in the night sky.

Here is a customized screen shot ready for viewing Orion and other constellations:

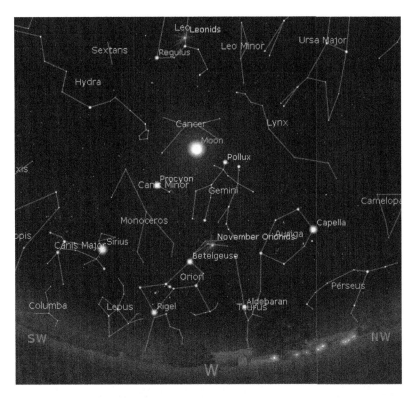

80 Customized Observation Screenshot (by Stellarium)

All you have to do is note that it is facing the western horizon and you do the same. You will find Orion just above the horizon, slightly to the left of West!

Orion is a good constellation to start with, as it is visible from any city in the world. How about other constellations? It depends on your Latitude. There is one blackout circular zone of known radius at that celestial pole which is further away from you than the other, otherwise the rest of the *CS* is visible over each 24-hour rotational period. For example, say your Latitude is 30°N. Thus, *CSP* is the celestial pole further away from you than the other. A circular zone around the *CSP* of radius = Latitude = 30° is blacked out to you, meaning you will not be able to observe any stars in it.

The boundary circle of this blackout zone has a declination of -60° (i.e. 90 minus 30, the negative sign indicating below the **CE**). Using it, you can consult the **Table of Coordinates: by Decl-RA** in **Appendix L: Information Tables** to see what stars of that table are blacked out: at least, all the stars from the top of the table up to Hadar (**Decl**-60°) will not be visible ('at least' because stars close to this horizon will likely also not be visible due to obstructions and / or atmospheric perturbation). The higher your Latitude the greater the circular blackout zone, maximum at the North or South Pole, minimum at the Equator. [Where I did my astronomy observations, back in 1972, at the University of Nairobi, it was ONE degree below the Equator. Heavenly! Pun intended.] So, at Latitude 30°N, you won't be able to see the nice, cute Southern Cross constellation called Crux, nor Triangulum Australe, nor Octans, ... As for observation Latitudes below the Equator, there is a similar circular blackout zone at the **CNP**.

Preparations for Finding Other Constellations

Once you have ascertained what is visible at your Latitude, that is outside the blackout zone, you can use the same approach as for Orion above to finding a suitable viewing time for your chosen celestial object. However, the above approach takes some time to do it. Here are a couple of suggestions to make the maximum return for your efforts:

- Instead of waiting for your desired constellation to transit, go with whatever currently transits. If you observe once a month, the transit time will change by two hours every month. Thus, in a couple of months, you will have observed star transits for 17% of the sky (2h X 2mths, divided by 24h, times 100). Better still, you will be covering a lot more than 17% because you can easily explore both sides of the transit by 2-3 hours of a transit, say an average of 2.5h each side of transit. Thus, to the 4h over two months above, you can add 2.5 X 2h, for both sides of transit: (4h + 5h, divided by 24h, times 100) = 38%. Thus, in less six months, you will have observed all the visible heavenly objects that appear in your night sky. [**Tip**: Keep a check-chart (akin to a checklist) in which you tick off or highlight the constellations observed. If nothing else, it can be motivating. Keep in mind that you will only be able to observe constellations outside of your blackout zone, so you would have to set your goals accordingly.]

- Widen your observation time window as you get better at it, so you get more and more over a shorter span of total elapsed time.

A Complete Observation Package

Here, I would like to share with you my observation package for my next visit to Tanga, Tanzania, my original hometown. It serves as a revision of observational steps previously presented, along with related commentary.

Site Information

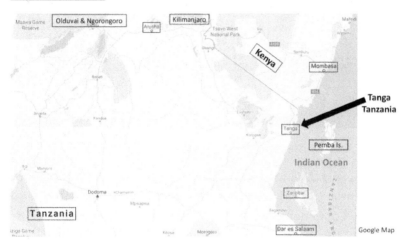

81 Tanga, Tanzania (Google Map)

Tanga is a coastal town in North-East Tanzania, an Indian Ocean port, just South of the Kenyan border.

Latitude: -5°. Longitude: 39° East of Greenwich.

Observation site: Ras Kazone promontory, East Cliff, facing East over the Indian Ocean.

Proposed date-time: 2019-07-07, 05:00:00.

Proposed sky segments to observe:

I. East, looking out to the Indian Ocean

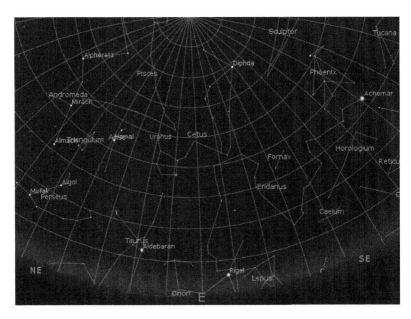

82 East Screenshot, Tanga (from Stellarium)

This screenshot is above the East horizon and up to the **Zenith** (the point at the top from where the vertical grid lines of the Local Horizon Coordinates emanate). The brightest star Rigel (7th) is close to the horizon and so, much affected by atmospheric attenuation, but I expect the no-light-pollution Indian Ocean to compensate and make it visible. It would be nice to trace the entire length of the River of constellation Eridanus. I expect the Andromeda Galaxy (M31) to be visible just off the Princess' waist, as seen in the zoom-in below:

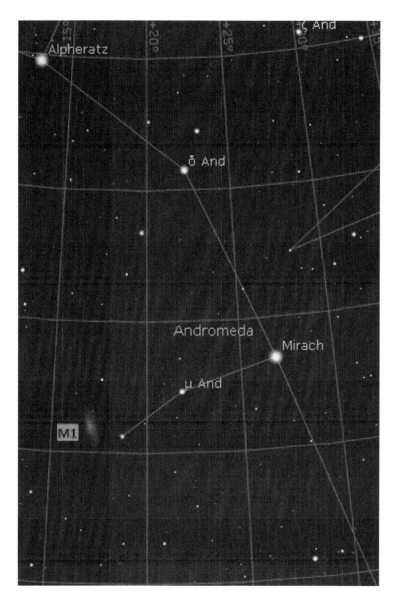

83 East View, Zoomed In (from Stellarium)

Steps to display the screenshot **#82** in Stellarium:

There are two sets of icons, one hidden below the bottom-left edge of the screen, the other off the lower-left edge. They are brought into view by nudging one of these edges with the mouse pointer.

1. Click the "Set normal time rate" icon in the bottom set of icons, to pause the clicking clock, so that the sky remains stationary.
2. Drag the default S(outh) horizon down to near bottom.
3. Click the "Azimuthal grid icon" in the bottom set of icons to display the *Azimuth-Altitude* grid lines. [Shortcut toggle: Z]
4. Click the "Location window icon" in the left set of icons and type in "Tanga". Click the "Tanga, Tanzania" entry in the results list. [Shortcut toggle: F6] Close the window.
5. Zoom out the sky with the mouse wheel. The horizon will rise up, so drag it back to near bottom. Keep doing these two steps until the *Zenith* comes into view near the top, as in the above screen shot #**82**. [Shortcut toggles: PgUp/PgDn for zooming. Up/down arrows ^/v for raising/lowering the horizon.]
6. Rotate the horizon by dragging it, until E(ast) is at the middle bottom. [Shortcut toggles: Left/right arrows </> for rotating the horizon.]
7. Click the "Date-time window" icon in the left set of icons and change the date and time to values given above. [Shortcut toggle: F6]
8. Only one bright star is seen: Achernar. So, click the "Sky viewing options" icon in the left set of icons and drag the "Labels and markers" slide bar to the right until more bright stars are labelled, as desired. See above screenshot. Close the window.
9. Click the "Constellation lines" icon in the left set of icons. [Shortcut toggle: C]

10. Click the "Constellation labels" icon in the left set of icons. [Shortcut toggle: V]
11. Voila!

Using the same process as above, the following screenshots were produced for the other cardinal points (North, West, South) as well as of the **Zenith**, making this a complete coverage of the sky.

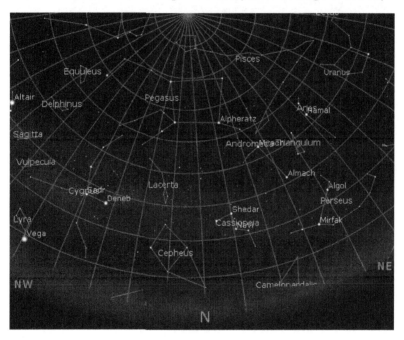

84 North Screenshot, Tanga (from Stellarium)

Nearly a dozen bright stars are seen above: Algol, Almach, Alpheratz, Altair, Deneb, Hamal, Mirach, Mirfak, Sadr, Shedar, Vegas.

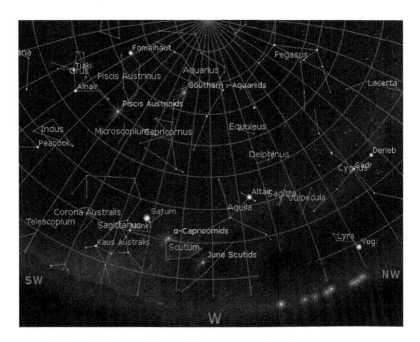

85 West Screenshot, Tanga (from Stellarium)

Just above the West horizon is the best view of the Milky Way band. You can see Stellarium's rendering of it as billowing darkly-glowing clouds from just below star Kaus Australis to star Deneb. The first of these stars is a marker to the Galactic Centre (more closely, the spout of Teapot asterism, below Kaus Australis)! The latter star marks Galactic Longitude 90° (approximately). This portion of the band is possibly its most glorious part.

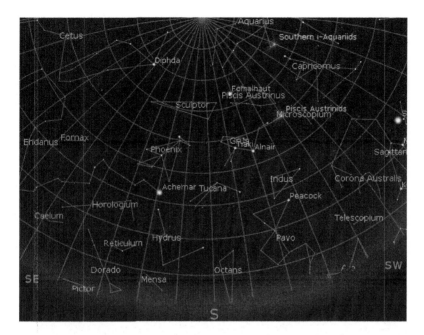

86 South Screenshot, Tanga (from Stellarium)

Three bright stars are seen in the above: Achernar, Diphda and Formalhaut.

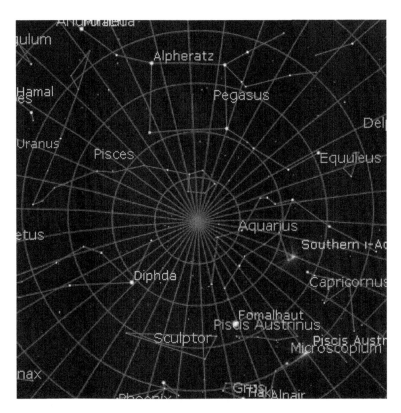

87 Zenith Screenshot up to Altitude50°, Tanga (from Stellarium)

Three bright stars are seen: Alpheratz, Diphda and Fomalhaut. The meteor shower Alpha Capricornids and Delta Aquariids (not shown above, but their constellations are) peak towards the end of July. Hopefully, I might see it even off-peak – refer to **Table of Near Galaxies, Meteor Showers & Nebulas** in **Appendix L: Information Tables**.

Appendix L: Information Tables

Table of Monthly Coordinates of the Sun

These are for the 1ˢᵗ of each month, at noon EST.

Month	RA	Decl
January	18.8h	-23°
February	21.0h	-17°
March	22.8h	-7°
VE	0h	0°
April	0.7h	5°
May	2.6h	15°
June	4.6h	22°
July	6.7h	23°
August	8.8h	18°
September	10.7h	8°
Fall Equinox	12h	0°
October	12.5h	-3°
November	14.4h	-15°
December	16.5h	-22°

Tables of Coordinates

The coordinates in the following tables were populated from the Simbad database at the University of Strasbourg, France, the values being rounded to the level of accuracy sufficient for naked eye astronomy. It contains at least the 300 brightest stars. Their brightness ranks are given in column titled "#". The 21 brightest are bolded, being the complete set of 1ˢᵗ Order *Magnitude* stars. Also, at least the brightest star of each OLD constellation is listed. Dim stars are given a ranking of '>333'. Where a common name does not exist, its Flamsteed number is used in the Common Name column, and if a Flamsteed number does not exist, it is left blank (hyphen) and the star therefore is identified by a Bayer

Greek-letter in the "Gk" column. [**Tip**: Brightness rankings may not be 100% consistent, especially when a star is actually a binary or multiple system e.g. two stars revolving around each other, thus its magnitude is either the sum of the two stars, or each is assigned a suffix such as A, B, each with its own magnitude. Generally, a multi-star system has been treated as originally observed, as a single star. The vast majority of the stars we see are binary stars, and many are multiple stars! For example, all of the stars of 1st Order *Magnitude*, except four, are binary or multiple star systems, mostly binary.] Another factor to keep in mind is, there are variable stars, meaning their *Magnitudes* vary over time. Having said all this, for our naked eye astronomy purposes, any reasonably good ranking is okay, perhaps grouped by orders of *Magnitude* (see order markers embedded in the table) instead of individual ranking e.g. the 21 stars in our *CE* model previously developed (see figure **22 Celestial Sphere showing the Milky Way band**) are a complete set of stars of first order *Magnitude*.]

These tables are provided here as a quick convenient reference. More precise numbers, of course, are to be found in databases like Simbad.

Several table versions are presented below:

Table of Coordinates: by Brightness

Table of Coordinates: by Common Name

Table of Coordinates: by Constellation

Table of Coordinates: by RA-Decl

The above table and the one below are useful in identifying an observed star whose Local Horizon Coordinates are converted to *RA* and *Decl* (see **Appendix C: Converting Local Horizon Coordinates to Celestial Coordinates**).

Table of Coordinates: by Decl-RA

Table of Coordinates: by Glat-Glong

The above table is useful for, inter alia, finding stars in the Milky Way band (a few degrees either side of Galactic Latitude 0°).

Following the above tables, is a **Table of Near Galaxies, Meteor Showers & Nebulas**.

A full list of constellations follows these tables at **Table of All Constellations, their Stories**. Below it, is a **Table of Asterisms, their Constellations**.

Table of Coordinates: by Brightness

is rank by brightness.

Superscript "NA" attached to a common name denotes "not approved" by the IAU.

Here are over 300 hundred named and unnamed stars (all named stars included), covering up to and including 5th Order **Magnitude**. As the eye gets accustomed to the dark sky and where light pollution is less (e.g. at higher **Altitudes**) and with experience, one should be able to easily see the 60+ stars of 2nd Order **Magnitude**, as well as 100± stars of 3rd Order **Magnitude**. In theory, one can see stars up to 6th Order **Magnitude**, but the higher the order, the better the viewing conditions it requires.

Common Name	#	Constellation	Gk	RA hrs	Decl. °	Glat. °	Glong. °
Gal. 0, 0		Sagittarius	-	17.8	-29	0	0
Gal. 0, 180		Taurus	-	5.8	29	0	180
Gal. 0, 270		Vela	-	9.2	-48	0	270
Gal. 0, 90		Cygnus	-	21.2	48	0	90
1ˢᵗ Order Mag.	0						
Sirius	1	Canis Major	α	6.8	-17	-9	227
Canopus	2	Carina	α	6.4	-53	-25	261
Rigel Kentaurus	3	Centaurus	α	14.7	-60	-1	316
Arcturus	4	Bootes	α	14.3	19	69	15
Vega	5	Lyra	α	18.6	38	19	67
Capella	6	Auriga	α	5.3	46	5	163
Rigel	7	Orion	α	5.2	-8	-25	209
Procyon	8	Canis Minor	α	7.7	5	13	214
Betelgeuse	9	Orion	β	5.9	7	-9	200
Achernar	10	Eridanus	α	1.6	-57	-59	291
Hadar	11	Centaurus	β	14.1	-60	1	312
Altair	12	Aquila	α	19.8	9	-9	48
Acrux	13	Crux	α	12.4	-63	-0	300
Aldebaran	14	Taurus	α	4.6	17	-20	181

Common Name	#	Constellation	Gk	RA hrs	Decl. °	Glat. °	Glong. °
Spica	15	Virgo	α	13.4	-11	51	316
Antares	16	Scorpius	α	16.5	-26	15	-8
Pollux	17	Gemini	β	7.8	28	23	192
Fomalhaut	18	Piscis Austrinus	α	23.0	-30	-65	21
Deneb	19	Cygnus	α	20.7	45	2.0	84
Mimosa	20	Crux	β	12.8	-60	33	303
Regulus	21	Leo	α	10.1	12	49	226
2nd Order Mag.	21.1						
Adhara	23	Canis Major	ε	7.0	-29	-11	240
Gacrux	24	Crux	γ	12.5	-57	6	300
Shaula	25	Scorpius	λ	17.6	-37	-2	352
Bellatrix	26	Orion	γ	5.4	6	-16	197
Elnath	27	Taurus	β	5.4	29	-4	178
Miaplacidus	28	Carina	β	9.2	-70	-14	286
Alnilam	29	Orion	ε	5.6	-1	-17	205
Alnair	30	Grus	α	22.1	-47	-53	350
Alnitak	31	Orion	ζ	5.7	-2	-17	207
Alioth	32	Ursa Major	ε	12.9	56	61	122
Kaus Australis	33	Sagittarius	ε	18.4	-34	-10	359
RegorNA	33	Vela	γ	8.2	-47	-8	263
Mirfak	34	Perseus	α	3.4	50	-6	147
Dubhe	35	Ursa Major	α	11.1	62	51	143
Wezen	36	Canis Major	δ	7.1	-26	-8	238
Alkaid / BenetnaschNA	37	Ursa Major	η	13.8	49	65	101
Avior	38	Carina	ε	8.4	-59	-13	274
Sargas	39	Scorpius	θ	17.7	-43	-6	347
Menkalinan	40	Auriga	β	6.0	45	10	167
Atria	41	Triangulum Australe	α	16.8	-69	-15	321
Alhena / AlmeisanNA	42	Gemini	γ	6.6	16	5	197
Peacock	43	Pavo	α	20.4	-57	-35	341
Polaris	44	Ursa Minor	α	2.5	89	27	123
Castor	45	Gemini	α	7.6	32	23	187
Mirzam	46	Canis Major	β	6.4	-18	-14	226
Alphard	47	Hydra	α	9.5	-9	29	241

Table of Coordinates: by Brightness

Common Name	#	Constellation	Gk	RA hrs	Decl. °	Glat. °	Glong. °
Alsephina	48	Vela	δ	8.7	-55	-7	272
Hamal	49	Aries	α	2.1	23	-36	145
Diphda	50	Cetus	β	0.7	-18	-81	111
Nunki	51	Sagittarius	σ	18.9	-26	-13	10
Menkent	52	Centaurus	θ	14.1	-36	24	319
Alpheratz	53	Andromeda	α	0.1	29	-33	112
Kochab	54	Ursa Minor	β	14.8	74	74	113
Mirach	55	Andromeda	β	1.2	36	-27	127
Saiph	56	Orion	κ	5.8	-10	-19	215
Rasalhague	57	Ophiuchus	α	17.6	13	23	36
Algol	58	Perseus	β	3.1	41	-15	3
Almach	59	Andromeda	γ	2.1	42	-19	137
Tiaki	60	Grus	β	22.7	-47	-58	346
Denebola	61	Leo	β	11.8	15	7`	251
Aspidiske / Turais^{NA}	62	Carina	ι	9.3	-59	-7	279
Muhlifain^{NA}	62	Centaurus	γ	12.7	-49	14	301
Naos	63	Puppis	ζ	8.1	-40	-5	256
Alphecca / Gemma^{NA}	64	Corona Borealis	α	15.6	27	54	42
Mizar	65	Ursa Major	ζ	13.4	55	62	113
Sadr	66	Cygnus	γ	20.4	40	2	78
Suhail	67	Vela	λ	9.1	-43	3	266
Eltanin	68	Draco	γ	17.9	51	29	79
Schedar	69	Cassiopeia	α	0.7	57	-6	121
Mintaka	70	Orion	δ	5.5	-0	-18	204
Caph	71	Cassiopeia	β	0.2	59	-3	117
Dschubba	72	Scorpius	δ	16.0	-23	23	350
Larawag	73	Scorpius	ε	16.8	-34	7	349
-	74	Centaurus	ε	13.7	-53	9	310
Merak	74	Ursa Major	β	11.0	56	55	149
Izar	75	Bootes	ε	14.7	27	65	39
Kakkab^{NA}	75	Lupus	α	14.7	-43	11	322
Enif	76	Pegasus	ε	21.7	10	-31	66
Ankaa	77	Phoenix	α	0.4	-42	-74	320
-	78	Centaurus	η	14.6	-42	17	323
Phecda	78	Urda Major	γ	11.6	54	61	141

Common Name	#	Constellation	Gk	RA hrs	Decl. °	Glat. °	Glong. °
Sabik	79	Ophiuchus	η	17.2	-16	14	7
Scheat	80	Pegasus	β	23.1	28	-29	96
Alderamin	81	Cepheus	α	21.3	63	9	101
Girtab[NA]	81	Scorpius	κ	17.6	-39	-5	351
Aludra	82	Canis Major	η	7.6	-29	-7	243
27	82	Cassiopeia	γ	0.9	61	-2	124
Markeb	83	Vela	κ	9.4	-55	-3	276
Aljanah / Gienah Cygni[NA]	84	Cygnus	ε	20.8	34	-6	76
Markab	85	Pegasus	α	23.1	15	-40	88
3rd Order Mag.	85.1						
Menkar	86	Cetus	α	3.0	4	-46	173
13	86	Ophiuchus	ζ	16.6	-11	24	6
Acrab / Graffias[NA]	87	Scorpius	β¹	16.1	-20	24	353
Zosma	88	Leo	δ	11.2	21	67	224
Arneb	89	Lepus	α	5.5	-18	-25	221
Gienah / Gienah Ghurab[NA]	90	Corvus	γ	12.3	-17	45	291
-	91	Centaurus	δ	12.1	-51	12	296
Ascella	91	Sagittarius	ζ	19.0	-30	-15	7
Algieba	92	Leo	γ¹	10.3	20	55	217
Zubene-schamali	93	Libra	β	15.3	-9	39	352
Unukalhai	94	Serpens	α	15.7	6	44	14
Sheratan	95	Aries	β	1.9	21	-40	142
Kraz	96	Corvus	β	12.6	-23	39	298
Mahasim	97	Auriga	θ	6.0	37	7	174
Phact	98	Columba	α	5.7	-34	-29	239
Ruchbah	99	Cassiopeia	δ	1.4	60	-2	127
Muphrid	100	Bootes	η	13.9	18	5	73
Hassaleh	101	Auriga	ι	5.0	33	-6	171
-	101	Musca	α	12.6	-69	-6	301
-	102	Vela	μ	11	-49	9	283
Lesath	102	Scorpius	υ	17.5	-37	-2	351
-	102	Puppis	π	7.3	-37	-11	249
Kaus Media	103	Sagittarius	δ	18.3	-29.8	-7.2	3.0

Table of Coordinates: by Brightness

Common Name	#	Constellation	Gk	RA hrs	Decl. °	Glat. °	Glong. °
Tarazed	104	Aquila	γ	19.8	11	-7	49
Athebyne / AlhibainNA	105	Draco	η	16.4	61	41	93
Yed Prior	106	Ophiuchus	δ	16.2	-3.7	32	9
Porrima	107	Virgo	γ	12.7	-1	61	298
-	107	Carina	θ	10.7	-64	-5	290
Imai	108	Crux	δ	12.3	-59	4	298
Zubenelgenubi	109	Libra	α²	14.8	-16	38	340
-	110	Centaurus	ι	13.3	-37	26	309
Cebalrai	110	Ophiuchus	β	17.6	5	17	29
Cursa / KursaNA	111	Eridanus	β	5.1	-5	-25	205
KekouanNA	111	Lupus	β	15.0	-43	14	326
Kornephoros	112	Hercules	β	16.5	21	40	39
Rastaban	113	Draco	β	17.5	52	33	80
Hatysa	114	Orion	ι	5.6	-6	-20	210
-	114	Lupus	γ	15.6	-41	12	333
Nihal	115	Lepus	β	5.5	-21	-27	224
Kaus Borealis	116	Sagittarius	λ	16.5	-25.4	-6.5	7.7
Paikauhale	117	Scorpius	τ	16.6	-28	13	352
-	117	Hydra	β	0.4	-77	-40	305
Algenib	118	Pegasus	γ	0.2	15	-47	109
Tureis	119	Puppis	ρ	8.1	-24	5	243
44	119	Perseus	ζ	3.9	32	-17	162
-	119	Ara	β	17.4	-55	-11	335
-	119	Ara	α	17.5	-50	-9	341
Alcyone	120	Taurus	η	3.8	24	-23	167
Deneb Algedi	121	Capricornus	δ	21.8	-16	-46	38
-	121	Triangulum Australe	β	15.9	-63	-7	322
Vindemaitrix	122	Virgo	ε	13.0	11	74	312
-	122	Hydrus	α	2.0	-62	54	289
Tejat	123	Gemini	μ	6.4	31	4	190
-	123	Tucana	α	22.3	-61	-48	330
-	123	Triangulum Australe	γ	15.3	-69	-9	316
Acamar	124	Eridanus	θ¹	3.0	-40	-61	248
Albaldah	125	Sagittarius	π	19.2	-21	-13	16

Common Name	#	Constellation	Gk	RA hrs	Decl. °	Glat. °	Glong. °
Alpherg	126	Pisces	η	1.5	45	-46	137
Cor Caroli	127	Canes Venatici	α²	12.9	38	79	118
Fang	128	Scorpius	π	16.0	-26	20	347
Gomeisa	129	Canis Minor	β	7.5	8	12	209
Alniyat	130	Scorpius	σ	16.4	-26	17	351
Fawaris	131	Cygnus	δ	19.7	45	10	79
Sadalsuud	132	Aquarius	β	21.5	-6	-38	48
45	133	Perseus	ε	4.0	40	-10	157
23	133	Perseus	γ	3.1	53	-4	142
-	133	Carina	υ	9.8	-65	-9	285
Matar	133	Pegasus	η	22.7	30	-25	93
Rutilicus^NA	133	Hercules	ζ	16.7	32	40	53
Algorab	134	Corvus	δ	12.5	-17	46	295
Sadalmelik	135	Aquarius	α	22.1	-1	-42	60
Tianguan	136	Taurus	ζ	5.6	21	-6	186
Zaurak	137	Eridanus	γ	4.0	-13	-45	205
17	137	Leo	ε	9.8	24	48	207
Alnasl	138	Sagittarius	γ²	18.1	-30	-5	1
Okab	139	Aquila	ζ	19.1	14	3	47
Aldhanab	140	Grus	γ	21.9	-37	-51	6
Pherkad	141	Ursa Minor	γ	15.3	72	41	108
Xamidimura	142	Scorpius	μ¹	16.9	-38	4	346
4	142	Triangulum	β	2.2	35	-25	141
Furud	143	Canis Major	ζ	6.3	-30	-19	238
Almaaz	144	Auriga	ε	5.0	44	1.2	163
Seginus	145	Bootes	γ	14.5	38	66	67
Albireo	146	Cygnus	β¹	19.5	28	5	62
Dabih	147	Capricornus	β¹	20.4	-15	-26	29
Mebsuta	148	Gemini	ε	6.7	25	10	189
Tania Australis	149	Ursa Major	μ	10.4	41	56	178
Altais	150	Draco	δ	19.2	40	23	99
Sarin	151	Hercules	δ	17.3	25	31	47
Talitha	152	Ursa Major	ι	9.0	48	41	172
40	153	Lynx	α	9.4	34	45	190
Wazn	153	Columba	β	5.8	-36	-27	241
67	154	Hercules	π	17.3	37	34	61
Aldhibah	154	Draco	ζ	17.1	66	35	96

Common Name	#	Constellation	Gk	RA hrs	Decl. °	Glat. °	Glong. °
Haedus	155	Auriga	η	5.1	41	0	165
Fuyue	156	Scorpius	η	5.1	14	0	165
Altaleban^NA	157	Puppis	τ	6.8	-51	-21	260
Tabit	157	Orion	π³	4.8	7	-23	191
Errai	158	Cepheus	γ	23.7	78	15	119
Alfirk	159	Cepheus	β	21.5	71	14	108
Yed Posterior	160	Ophiuchus	ε	16.3	-4.7	31	9
65	161	Aquila	θ	20.2	-1	-18	42
Brachium	161	Libra	σ	15.1	-25	29	337
Sulafat	162	Lyra	γ	19.0	33	13	63
Skat	163	Aquarius	δ	22.9	-16	-61	50
Edasich	164	Draco	ι	15.4	59	49	94
Propus	165	Gemini	η	16.2	23	3	189
Megrez	166	Ursa Major	δ	12.3	57	59	133
Chertan	167	Leo	θ	11.2	15.4	65	235
Alzirr	168	Gemini	ξ	6.8	13	5	201
Muscida	169	Ursa Major	o	8.5	61	35	156
Segin	170	Cassiopeia	ε	1.9	64	2	130
Rasalgethi	171	Hercules	α¹	17.2	14	28	36
Heze	172	Virgo	ζ	13.6	-1	60	325
Meissa / Heka^NA	173	Orion	λ	5.6	10	-12	195
Minelauva	174	Virgo	δ	12.9	3	66	306
Homam	175	Pegasus	ζ	22.7	11	-41	79
Mothallah	176	Triangulum	α	1.9	30	-31	139
Adhafera	177	Leo	ζ	10.3	23	55	210
Azmidi	178	Puppis	ξ	7.8	-25	1	241
Tania Borealis	179	Ursa Major	λ	10.3	43	55	176
49	180	Bootes	δ	15.3	33	58	53
Achird	180	Cassiopeia	η	0.8	58	-5	123
Alula Borealis	182	Ursa Major	ν	11.3	33	69	191
Ashlesha	183	Hydra	ε	8.8	6	29	221
Nekkar	183	Bootes	β	15.0	40	60	68
Unurgunite	184	Canis Major	σ	7.0	-28	-10	239
4th Order Mag.	184.1						
Wasat	185	Gemini	δ	7.3	22	16	196
Sadalbari	186	Pegasus	μ	22.8	25	-31	91
Biham	187	Pegasus	θ	22.2	6	-39	67

Table of Coordinates: by Brightness

Common Name	#	Constellation	Gk	RA hrs	Decl. °	Glat. °	Glong. °
Sheliak	188	Lyra	β	18.8	33	15	63
Subra	189	Leo	o	9.7	10	42	225
Ain	190	Taurus	ε	4.5	19	-20	178
Tarf / Altarf^{NA}	191	Cancer	β	8.3	9	23	214
Kaffaljidhma	192	Ceteus	γ	2.7	14	-49	169
Pipirima	193	Scorpius	μ²	16.9	-38	4	346
Algedi	194	Capricornus	α²	20.3	-13	-25	31
Ginan	195	Crux	ε	12.4	-60	2	299
Nembus / 51 / υ Persesu	196	Andromeda	-	1.6	49	-14	131
Zavijava	197	Virgo	β	11.8	46	61	271
Bharani / 41	198	Aries	-	2.8	27	-29	153
Atlas / 27	199	Taurus	-	3.8	24	-23	167
Rotanev	200	Delphinus	β	20.6	15	-16	59
Prima Hyadum	201	Taurus	γ	4.3	16	-24	179
Nusakan	202	Corona Borealist	β	15.5	29	56	46
Thuban	203	Draco	α	14.1	64	51	111
Fulu	204	Cassiopeia	ζ	0.6	54	-9	121
Nashira	205	Capricornus	γ	21.7	-17	-45	36
Saclateni	206	Auriga	ζ	5.0	41	-0	165
Taiyangshou	207	Ursa Major	χ	11.8	48	66	150
Alshain	208	Aquila	β	19.9	6	-11	46
Electra / 17	209	Taurus	-	3.7	24	-24	166
Chamukuy	210	Taurus	θ	4.5	16	-22	180
Grumium	211	Draco	ξ	17.9	57	30	85
Ran	212	Eridanus	ε	3.5	-9	-48	196
Baten Kaitos	213	Ceteus	ζ	1.9	-10	-68	166
Miram	214	Perseus	η	2.8	56	-3	138
Sualocin	215	Delphinus	α	20.7	16	-15	60
Albali	216	Aquarius	ε	20.8	-9	-30	38
Secunda Hyadum	217	Taurus	δ	4.4	18	-22	178
Misam	218	Perseus	κ	3.2	45	-11	147
Praecipua / 46	219	Leo Minor	-	10.9	34	64	190
Theemin	220	Eridinus	υ	4.6	-31	-41	231
Alrescha	221	Pisces	α	2.0	3	-56	155

Table of Coordinates: by Brightness

Common Name	#	Constellation	Gk	RA hrs	Decl. °	Glat. °	Glong. °
Giausar	222	Draco	λ	11.5	69	+46	133
Marfik	223	Opheuchus	λ	16.5	2	32	17
Atik	224	Perseus	o	3.7	32	-18	160
Polis	225	Sagittarius	μ	18.2	-21	-2	10
Dalim / 12 Eridanus	226	Fornax	α	3.2	-29	-59	225
Sadachbia	227	Aquarius	γ	22.4	-1	-46	62
Iklil	228	Scorpius	ρ	15.9	-29	18	345
Maia / 20	229	Taurus	-	3.8	24	-24	166
Rasalas	230	Leo	μ	9.9	26	50	204
Azha	231	Eridanus	η	2.9	-8.9	-55	187
Zaniah	232	Virgo	η	12.3	-1	61	286
Ukdah	233	Hydra	ι	9.7	-1	36	236
Zubenelhakradi	234	Libra	γ	15.6	-15	32	352
Kitalpha	235	Equuleus	α	21.3	5	-29	56
Asellus Asutralis	236	Cancer	δ	8.7	18	33	208
Arkab Prior	237	Sagittarius	β¹	19.4	-44	-24	354
Rukbat	238	Sagittarius	α	19.4	-41	-23	358
Beemim	239	Eridanus	υ³	4.4	-34	-44	235
Menkib	240	Perseus	ξ	4.0	36	-13	160
Alcor	241	Ursa Major	γ	13.6	55	61	113
Mekbuda	242	Gemini	ζ	7.1	21	12	196
Alchiba	243	Corvus	α	12.1	-25	37	291
Sceptrum / 53	244	Eridanus	-	4.6	-14	-36	211
Wurren	245	Phoenix	ζ	1.1	-55	-62	298
Deneb El Okab[NA]	245	Aquila	ε	19.0	15	5	47
Aldulfin	246	Delphinus	ε	20.6	11	-17	55
Beid	247	Eridanus	σ¹	4.2	-7	-38	199
Syrma	248	Virgo	ι	14.3	-6	51	338
Alkes	249	Crater	α	11	-18	37	269
Gudja	250	Serpens	κ	15.6	18	48	30
Titawin	251	Andromeda	υ	1.6	41	-21	132
Meridiana	252	Corona Australis	α	19.2	-38	-20	359
Muliphein	253	Canis Major	γ	7.1	-16	-4	228
Zhang	254	Hydra	υ¹	9.9	-15	29	251

Table of Coordinates: by Brightness

Common Name	#	Constellation	Gk	RA hrs	Decl. °	Glat. °	Glong. °
Merope / 23	255	Taurus	-	3.8	24	-24	167
Alkaphrah	256	Ursa Major	κ	9.1	47	42	173
Ancha	257	Aquarius	θ	22.3	47	-49	54
Kang	258	Virgo	κ	14.2	-10	48	334
Xuange	259	Bootes	λ	14.3	46	65	87
Chara	260	Canes Venatici	β	12.6	41	75	136
Alsciaukat / 31	261	Lynx	-	8.4	43	34	177
Acubens	262	Cancer	α	9.0	12	34	217
Kurhah	263	Cepheus	ξ	22.1	38	7	106
Arkab Posterior	264	Sagittarius	β²	19.4	-45	-24	353
Torcular	265	Pisces	o	1.8	9	-51	145
Taygeta / 19	266	Taurus	q	3.8	24	-24	166
Alkalurops	267	Bootes	μ²	15.4	37	56	60
Alterf	269	Leo	λ	9.5	23	45	207
Botein	269	Aries	δ	3.2	20	-32	163
Yildun	270	Ursa Minor	δ	17.5	87	28	119
Sham	271	Sagitta	α	19.6	18	-2	55
Alula Australis	272	Ursa Major	ξ	11.3	32	69	195
Maasym	273	Hercules	λ	17.5	26	28	49
Alkarab	274	Pegasus	υ	23.4	23	-35	99
Aladfar	275	Lyra	η	19.2	39	13	71
Keid	276	Eridanus	o	4.3	-8	-38	201
Anser	277	Vulcan	α	19.5	25	3	59
Minchir	278	Hydra	σ	8.6	3	25	223
Fumalsamakah	279	Pisces	β	23.1	4	-50	79
5th Order Mag.	279.1						
Jabbah	280	Scorpius	ν	16.2	-19	23	355
Khambalia	281	Virgo	λ	14.3	-13	44	333
Lilii Borea / 39	282	Aries	-	2.8	29	-27	151
Cujam	283	Hercules	ω	16.4	14	39	29
Dziban	284	Draco	ψ¹	17.7	72	31	103
Salm	285	Pegasus	τ	23.3	24	-35	98
Alya	286	Serpens	θ¹	18.9	4.2	1	37
Castula	287	Cassiopeia	υ²	0.9	59	-4	124
Asellus Borealis	288	Cancer	γ	8.7	21	34	204
Alsafi	289	Draco	σ	19.5	70	22	101
Tegmine	290	Cancer	ζ¹	8.2	18	26	205

Table of Coordinates: by Brightness   **187** | P a g e

Common Name	#	Constellation	Gk	RA hrs	Decl. °	Glat. °	Glong. °
Azelfafage	291	Cygnus	π^1	21.7	51	-1	95
Terebellum	292	Sagittarius	ω	19.9	-26	-25	15
Libertas	293	Aquila	ξ	19.9	8	-10	48
Mesarthim	294	Aries	γ^1	1.9	19	-41	143
Angetenar	295	Eridanus	τ^2	2.9	-21	-62	207
Alshat	296	Capricornus	ν	20.3	-13	-25	31
Bunda	297	Aquarius	ξ	21.6	-8	-40	46
Zibal	298	Eridanus	ζ	3.3	-9	-51	192
Fafnir / 42	299	Draco	-	18.4	66	27	95
Elgafar	300	Virgo	φ	14.5	-2	53	345
Diadem	301	Coma Berenices	α	3.2	18	79	328
Ainalrami	302	Sagittarius	ν	18.9	-23	-11	13
Adhil	303	Andromeda	ξ	1.4	46	-17	129
Jishui	304	Gemini	o	7.7	35	24	185
Felis	305	Hydra	-	9.9	-19	27	255
Elkurud	306	Columba	θ	6.1	-37	-24	244
Marsic	307	Hercules	κ	16.1	17	44	31
Chalawan / 47	308	Ursa Major	-	11.0	40	63	176
Situla	309	Aquarius	κ	22.6	-4	-51	63
Pleione / 28	310	Taurus	-	3.8	24	-23	167
Guniibuu / 36	311	Ophiuchus	-	17.3	-27	7	358
Cervantes	312	Ara	μ	17.7	-52	-11	341
Revati	313	Pisces	ζ	1.2	8	-55	133
Veritate / 14	314	Andromeda	-	23.5	39	-21	106
Taiyi / 8	315	Draco	-	12.9	65	52	122
Intercrus / 41 LynxNA	316	Ursa Major	-	9.5	46	46	174
La Superba	317	Canes Venatici	γ	12.8	45	72	127
Tianyi / 7	318	Draco	-	12.8	67	50	124
Celaeno / 16	319	Taurus	-	3.7	24	-24	166
Polaris Australis	320	Octans	σ	21.1	-89	-28	304
Musica / 18	321	Delphinus	-	21.0	11	-22	59
Helvetios / 51	322	Pegasus	-	23.0	21	-35	90
6th Order Mag.	322.1						
Alrakis	323	Draco	μ	17.1	54	37	82
Nahn	324	Cancer	ξ	9.2	22	40	206

Table of Coordinates: by Brightness   **188** | P a g e

Common Name	#	Constellation	Gk	RA hrs	Decl. °	Glat. °	Glong. °
Alruba	325	Draco	-	17.7	54	31	81
Asterope / 21	326	Taurus	-	3.8	25	-23	166
Merga / 38	327	Bootes	-	14.8	46	60	80
Tonatiuh	328	Camelopardalis	-	12.1	77	40	126
Piautos	329	Cancer	λ	8.3	24	30	199
Copernicus / 55	330	Cancer	-	8.9	28	38	197
Meleph	331	Cancer	ε	8.7	20	32	206
Mira	332	Ceteus	o	2.3	-3	-58	168
>6th Order Mag.	333						
107	>333	Virgo	μ	14.7	-6	47	347
109	>333	Virgo	-	14.8	2	53	355
24	>333	Libra	ι	15.2	-20	32	343
30	>333	Bootes	ζ	14.7	14	61	11
37	>333	Aries	o	2.7	15	-39	159
38	>333	Lynx	-	9.3	37	45	187
55	>333	Aquila	η	19.9	1	-13	41
63	>333	Cygnus	-	21.1	48	0	89
85	>333	Hercules	ι	17.7	46	31	72
94	>333	Taurus	τ	4.7	23	-15	177
HIP 4363	>333	Sculptor	-	0.9	-27	-89	114
HIP 62752 A	>333	Coma Berenices	-	12.9	27	90	120

Table of Coordinates: by Common Name

Common Name	#	Constellation	Gk	RA hrs	Decl. °	Glat. °	Glong. °
-	74	Centaurus	ε	13.7	-53	9	310
-	78	Centaurus	η	14.6	-42	17	323
-	91	Centaurus	δ	12.1	-51	12	296
-	101	Musca	α	12.6	-69	-6	301
-	102	Vela	μ	11	-49	9	283
-	102	Puppis	π	7.3	-37	-11	249
-	107	Carina	θ	10.7	-64	-5	290
-	110	Centaurus	ι	13.3	-37	26	309
-	114	Lupus	γ	15.6	-41	12	333
-	117	Hydra	β	0.4	-77	-40	305
-	119	Ara	β	17.4	-55	-11	335
-	119	Ara	α	17.5	-50	-9	341
-	121	Triangulum Australe	β	15.9	-63	-7	322
-	122	Hydrus	α	2.0	-62	54	289
-	123	Tucana	α	22.3	-61	-48	330
-	123	Triangulum Australe	γ	15.3	-69	-9	316
-	133	Carina	υ	9.8	-65	-9	285
107	>333	Virgo	μ	14.7	-6	47	347
109	>333	Virgo	-	14.8	2	53	355
13	86	Ophiuchus	ζ	16.6	-11	24	6
17	137	Leo	ε	9.8	24	48	207
23	133	Perseus	γ	3.1	53	-4	142
24	>333	Libra	ι	15.2	-20	32	343
30	>333	Bootes	ζ	14.7	14	61	11
37	>333	Aries	o	2.7	15	-39	159
38	>333	Lynx	-	9.3	37	45	187
4	142	Triangulum	β	2.2	35	-25	141
40	153	Lynx	α	9.4	34	45	190
44	119	Perseus	ζ	3.9	32	-17	162
45	133	Perseus	ε	4.0	40	-10	157

Common Name	#	Constellation	Gk	RA hrs	Decl. °	Glat. °	Glong. °
49	180	Bootes	δ	15.3	33	58	53
55	>333	Aquila	η	19.9	1	-13	41
63	>333	Cygnus	-	21.1	48	0	89
65	161	Aquila	θ	20.2	-1	-18	42
67	154	Hercules	π	17.3	37	34	61
85	>333	Hercules	ι	17.7	46	31	72
94	>333	Taurus	τ	4.7	23	-15	177
Acamar	124	Eridanus	θ¹	3.0	-40	-61	248
Achernar	10	Eridanus	α	1.6	-57	-59	291
Achird	180	Cassiopeia	η	0.8	58	-5	123
Acrab / GraffiasNA	87	Scorpius	β¹	16.1	-20	24	353
Acrux	13	Crux	α	12.4	-63	-0	300
Acubens	262	Cancer	α	9.0	12	34	217
Adhafera	177	Leo	ζ	10.3	23	55	210
Adhara	23	Canis Major	ε	7.0	-29	-11	240
Adhil	303	Andromeda	ξ	1.4	46	-17	129
Ain	190	Taurus	ε	4.5	19	-20	178
Ainalrami	302	Sagittarius	ν	18.9	-23	-11	13
Aladfar	275	Lyra	η	19.2	39	13	71
Albaldah	125	Sagittarius	π	19.2	-21	-13	16
Albali	216	Aquarius	ε	20.8	-9	-30	38
Albireo	146	Cygnus	β¹	19.5	28	5	62
Alchiba	243	Corvus	α	12.1	-25	37	291
Alcor	241	Ursa Major	γ	13.6	55	61	113
Alcyone	120	Taurus	η	3.8	24	-23	167
Aldebaran	14	Taurus	α	4.6	17	-20	181
Alderamin	81	Cepheus	α	21.3	63	9	101
Aldhanab	140	Grus	γ	21.9	-37	-51	6
Aldhibah	154	Draco	ζ	17.1	66	35	96
Aldulfin	246	Delphinus	ε	20.6	11	-17	55
Alfirk	159	Cepheus	β	21.5	71	14	108
Algedi	194	Capricornus	α²	20.3	-13	-25	31
Algenib	118	Pegasus	γ	0.2	15	-47	109
Algieba	92	Leo	γ¹	10.3	20	55	217
Algol	58	Perseus	β	3.1	41	-15	3
Algorab	134	Corvus	δ	12.5	-17	46	295

Table of Coordinates: by Common Name   **191** | P a g e

Common Name	#	Constellation	Gk	RA hrs	Decl. °	Glat. °	Glong. °
Alhena / Almeisan^NA	42	Gemini	γ	6.6	16	5	197
Alioth	32	Ursa Major	ε	12.9	56	61	122
Aljanah / Gienah Cygni^NA	84	Cygnus	ε	20.8	34	-6	76
Alkaid / Benetnasch^NA	37	Ursa Major	η	13.8	49	65	101
Alkalurops	267	Bootes	μ²	15.4	37	56	60
Alkaphrah	256	Ursa Major	κ	9.1	47	42	173
Alkarab	274	Pegasus	υ	23.4	23	-35	99
Alkes	249	Crater	α	11	-18	37	269
Almaaz	144	Auriga	ε	5.0	44	1.2	163
Almach	59	Andromeda	γ	2.1	42	-19	137
Alnair	30	Grus	α	22.1	-47	-53	350
Alnasl	138	Sagittarius	γ²	18.1	-30	-5	1
Alnilam	29	Orion	ε	5.6	-1	-17	205
Alnitak	31	Orion	ζ	5.7	-2	-17	207
Alniyat	130	Scorpius	σ	16.4	-26	17	351
Alphard	47	Hydra	α	9.5	-9	29	241
Alphecca / Gemma^NA	64	Corona Borealis	α	15.6	27	54	42
Alpheratz	53	Andromeda	α	0.1	29	-33	112
Alpherg	126	Pisces	η	1.5	45	-46	137
Alrakis	323	Draco	μ	17.1	54	37	82
Alrescha	221	Pisces	α	2.0	3	-56	155
Alruba	325	Draco	-	17.7	54	31	81
Alsafi	289	Draco	σ	19.5	70	22	101
Alsciaukat / 31	261	Lynx	-	8.4	43	34	177
Alsephina	48	Vela	δ	8.7	-55	-7	272
Alshain	208	Aquila	β	19.9	6	-11	46
Alshat	296	Capricornus	ν	20.3	-13	-25	31
Altair	12	Aquila	α	19.8	9	-9	48
Altais	150	Draco	δ	19.2	40	23	99
Altaleban^NA	157	Puppis	τ	6.8	-51	-21	260
Alterf	269	Leo	λ	9.5	23	45	207
Aludra	82	Canis Major	η	7.6	-29	-7	243
Alula Australis	272	Ursa Major	ξ	11.3	32	69	195

Table of Coordinates: by Common Name   **192** | P a g e

Common Name	#	Constellation	Gk	RA hrs	Decl. °	Glat. °	Glong. °
Alula Borealis	182	Ursa Major	ν	11.3	33	69	191
Alya	286	Serpens	θ¹	18.9	4.2	1	37
Alzirr	168	Gemini	ξ	6.8	13	5	201
Ancha	257	Aquarius	θ	22.3	47	-49	54
Angetenar	295	Eridanus	τ²	2.9	-21	-62	207
Ankaa	77	Phoenix	α	0.4	-42	-74	320
Anser	277	Vulcan	α	19.5	25	3	59
Antares	16	Scorpius	α	16.5	-26	15	-8
Arcturus	4	Bootes	α	14.3	19	69	15
Arkab Posterior	264	Sagittarius	β²	19.4	-45	-24	353
Arkab Prior	237	Sagittarius	β¹	19.4	-44	-24	354
Arneb	89	Lepus	α	5.5	-18	-25	221
Ascella	91	Sagittarius	ζ	19.0	-30	-15	7
Asellus Asutralis	236	Cancer	δ	8.7	18	33	208
Asellus Borealis	288	Cancer	γ	8.7	21	34	204
Ashlesha	183	Hydra	ε	8.8	6	29	221
Aspidiske / TuraisNA	62	Carina	ι	9.3	-59	-7	279
Asterope / 21	326	Taurus	-	3.8	25	-23	166
Athebyne / AlhibainNA	105	Draco	η	16.4	61	41	93
Atik	224	Perseus	o	3.7	32	-18	160
Atlas / 27	199	Taurus	-	3.8	24	-23	167
Atria	41	Triangulum Australe	α	16.8	-69	-15	321
Avior	38	Carina	ε	8.4	-59	-13	274
Azelfafage	291	Cygnus	π¹	21.7	51	-1	95
Azha	231	Eridanus	η	2.9	-8.9	-55	187
Azmidi	178	Puppis	ξ	7.8	-25	1	241
Baten Kaitos	213	Ceteus	ζ	1.9	-10	-68	166
Beemim	239	Eridanus	υ³	4.4	-34	-44	235
Beid	247	Eridanus	σ¹	4.2	-7	-38	199
Bellatrix	26	Orion	γ	5.4	6	-16	197
Betelgeuse	9	Orion	β	5.9	7	-9	200
Bharani / 41	198	Aries	-	2.8	27	-29	153
Biham	187	Pegasus	θ	22.2	6	-39	67
Botein	269	Aries	δ	3.2	20	-32	163

Common Name	#	Constellation	Gk	RA hrs	Decl. °	Glat. °	Glong. °
Brachium	161	Libra	σ	15.1	-25	29	337
Bunda	297	Aquarius	ξ	21.6	-8	-40	46
Canopus	2	Carina	α	6.4	-53	-25	261
Capella	6	Auriga	α	5.3	46	5	163
Caph	71	Cassiopeia	β	0.2	59	-3	117
Castor	45	Gemini	α	7.6	32	23	187
Castula	287	Cassiopeia	υ²	0.9	59	-4	124
Cebalrai	110	Ophiuchus	β	17.6	5	17	29
Celaeno / 16	319	Taurus	-	3.7	24	-24	166
Cervantes	312	Ara	μ	17.7	-52	-11	341
Chalawan / 47	308	Ursa Major	-	11.0	40	63	176
Chamukuy	210	Taurus	θ	4.5	16	-22	180
Chara	260	Canes Venatici	β	12.6	41	75	136
Chertan	167	Leo	θ	11.2	15.4	65	235
Copernicus / 55	330	Cancer	-	8.9	28	38	197
Cor Caroli	127	Canes Venatici	α²	12.9	38	79	118
Cujam	283	Hercules	ω	16.4	14	39	29
Cursa / Kursa[NA]	111	Eridanus	β	5.1	-5	-25	205
Dabih	147	Capricornus	β¹	20.4	-15	-26	29
Dalim / 12 Eridanus	226	Fornax	α	3.2	-29	-59	225
Deneb	19	Cygnus	α	20.7	45	2.0	84
Deneb Algedi	121	Capricornus	δ	21.8	-16	-46	38
Deneb El Okab[NA]	245	Aquila	ε	19.0	15	5	47
Denebola	61	Leo	β	11.8	15	7`	251
Diadem	301	Coma Berenices	α	3.2	18	79	328
Diphda	50	Cetus	β	0.7	-18	-81	111
Dschubba	72	Scorpius	δ	16.0	-23	23	350
Dubhe	35	Ursa Major	α	11.1	62	51	143
Dziban	284	Draco	ψ¹	17.7	72	31	103
Edasich	164	Draco	ι	15.4	59	49	94
Electra / 17	209	Taurus	-	3.7	24	-24	166
Elgafar	300	Virgo	φ	14.5	-2	53	345
Elkurud	306	Columba	θ	6.1	-37	-24	244
Elnath	27	Taurus	β	5.4	29	-4	178

Common Name	#	Constellation	Gk	RA hrs	Decl. °	Glat. °	Glong. °
Eltanin	68	Draco	γ	17.9	51	29	79
Enif	76	Pegasus	ε	21.7	10	-31	66
Errai	158	Cepheus	γ	23.7	78	15	119
Fafnir / 42	299	Draco	-	18.4	66	27	95
Fang	128	Scorpius	π	16.0	-26	20	347
Fawaris	131	Cygnus	δ	19.7	45	10	79
Felis	305	Hydra	-	9.9	-19	27	255
Fomalhaut	18	Piscis Austrinus	α	23.0	-30	-65	21
Fulu	204	Cassiopeia	ζ	0.6	54	-9	121
Fumalsamakah	279	Pisces	β	23.1	4	-50	79
Furud	143	Canis Major	ζ	6.3	-30	-19	238
Fuyue	156	Scorpius	η	5.1	14	0	165
Gacrux	24	Crux	γ	12.5	-57	6	300
Gal. 0, 0		Sagittarius	-	17.8	-29	0	0
Gal. 0, 180		Taurus	-	5.8	29	0	180
Gal. 0, 270		Vela	-	9.2	-48	0	270
Gal. 0, 90		Cygnus	-	21.2	48	0	90
Giausar	222	Draco	λ	11.5	69	+46	133
Gienah / Gienah GhurabNA	90	Corvus	γ	12.3	-17	45	291
Ginan	195	Crux	ε	12.4	-60	2	299
GirtabNA	81	Scorpius	κ	17.6	-39	-5	351
Gomeisa	129	Canis Minor	β	7.5	8	12	209
Grumium	211	Draco	ξ	17.9	57	30	85
Gudja	250	Serpens	κ	15.6	18	48	30
Guniibuu / 36	311	Ophiuchus	-	17.3	-27	7	358
Hadar	11	Centaurus	β	14.1	-60	1	312
Haedus	155	Auriga	η	5.1	41	0	165
Hamal	49	Aries	α	2.1	23	-36	145
Hassaleh	101	Auriga	ι	5.0	33	-6	171
Hatysa	114	Orion	ι	5.6	-6	-20	210
Helvetios / 51	322	Pegasus	-	23.0	21	-35	90
Heze	172	Virgo	ζ	13.6	-1	60	325
HIP 4363	>333	Sculptor	-	0.9	-27	-89	114
HIP 62752 A	>333	Coma Berenices	-	12.9	27	90	120
Homam	175	Pegasus	ζ	22.7	11	-41	79

Table of Coordinates: by Common Name   **195** | P a g e

Common Name	#	Constellation	Gk	RA hrs	Decl. °	Glat. °	Glong. °
Iklil	228	Scorpius	ρ	15.9	-29	18	345
Imai	108	Crux	δ	12.3	-59	4	298
Intercrus / 41 Lynx[NA]	316	Ursa Major	-	9.5	46	46	174
Izar	75	Bootes	ε	14.7	27	65	39
Jabbah	280	Scorpius	ν	16.2	-19	23	355
Jishui	304	Gemini	o	7.7	35	24	185
Kaffaljidhma	192	Ceteus	γ	2.7	14	-49	169
Kakkab[NA]	75	Lupus	α	14.7	-43	11	322
Kang	258	Virgo	κ	14.2	-10	48	334
Kaus Australis	33	Sagittarius	ε	18.4	-34	-10	359
Kaus Borealis	116	Sagittarius	λ	16.5	-25.4	-6.5	7.7
Kaus Media	103	Sagittarius	δ	18.3	-29.8	-7.2	3.0
Keid	276	Eridanus	o	4.3	-8	-38	201
Kekouan[NA]	111	Lupus	β	15.0	-43	14	326
Khambalia	281	Virgo	λ	14.3	-13	44	333
Kitalpha	235	Equulus	α	21.3	5	-29	56
Kochab	54	Ursa Minor	β	14.8	74	74	113
Kornephoros	112	Hercules	β	16.5	21	40	39
Kraz	96	Corvus	β	12.6	-23	39	298
Kurhah	263	Cepheus	ξ	22.1	38	7	106
La Superba	317	Canes Venatici	γ	12.8	45	72	127
Larawag	73	Scorpius	ε	16.8	-34	7	349
Lesath	102	Scorpius	υ	17.5	-37	-2	351
Libertas	293	Aquila	ξ	19.9	8	-10	48
Lilii Borea / 39	282	Aries	-	2.8	29	-27	151
Maasym	273	Hercules	λ	17.5	26	28	49
Mahasim	97	Auriga	θ	6.0	37	7	174
Maia / 20	229	Taurus	-	3.8	24	-24	166
Marfik	223	Opheuchus	λ	16.5	2	32	17
Markab	85	Pegasus	α	23.1	15	-40	88
Markeb	83	Vela	κ	9.4	-55	-3	276
Marsic	307	Hercules	κ	16.1	17	44	31
Matar	133	Pegasus	η	22.7	30	-25	93
Mebsuta	148	Gemini	ε	6.7	25	10	189
Megrez	166	Ursa Major	δ	12.3	57	59	133
Meissa / Heka[NA]	173	Orion	λ	5.6	10	-12	195

Common Name	#	Constellation	Gk	RA hrs	Decl. °	Glat. °	Glong. °
Mekbuda	242	Gemini	ζ	7.1	21	12	196
Meleph	331	Cancer	ε	8.7	20	32	206
Menkalinan	40	Auriga	β	6.0	45	10	167
Menkar	86	Cetus	α	3.0	4.1	-46	173
Menkent	52	Centaurus	θ	14.1	-36	24	319
Menkib	240	Perseus	ξ	4.0	36	-13	160
Merak	74	Ursa Major	β	11.0	56	55	149
Merga / 38	327	Bootes	-	14.8	46	60	80
Meridiana	252	Corona Australis	α	19.2	-38	-20	359
Merope / 23	255	Taurus	-	3.8	24	-24	167
Mesarthim	294	Aries	γ¹	1.9	19	-41	143
Miaplacidus	28	Carina	β	9.2	-70	-14	286
Mimosa	20	Crux	β	12.8	-60	33	303
Minchir	278	Hydra	σ	8.6	3	25	223
Minelauva	174	Virgo	δ	12.9	3	66	306
Mintaka	70	Orion	δ	5.5	-0	-18	204
Mira	332	Ceteus	o	2.3	-3	-58	168
Mirach	55	Andromeda	β	1.2	36	-27	127
Miram	214	Perseus	η	2.8	56	-3	138
Mirfak	34	Perseus	α	3.4	50	-6	147
Mirzam	46	Canis Major	β	6.4	-18	-14	226
Misam	218	Perseus	κ	3.2	45	-11	147
Mizar	65	Ursa Major	ζ	13.4	55	62	113
Mothallah	176	Triangulum	α	1.9	30	-31	139
Muhlifain[NA]	62	Centaurus	γ	12.7	-49	14	301
Muliphein	253	Canis Major	γ	7.1	-16	-4	228
Muphrid	100	Bootes	η	13.9	18	5	73
Muscida	169	Ursa Major	o	8.5	61	35	156
Musica / 18	321	Delphinus	-	21.0	11	-22	59
Nahn	324	Cancer	ξ	9.2	22	40	206
Naos	63	Puppis	ζ	8.1	-40	-5	256
Nashira	205	Capricornus	γ	21.7	-17	-45	36
Nekkar	183	Bootes	β	15.0	40	60	68
Nembus / 51 / υ Persesu	196	Andromeda	-	1.6	49	-14	131
Nihal	115	Lepus	β	5.5	-21	-27	224

Common Name	#	Constellation	Gk	RA hrs	Decl. °	Glat. °	Glong. °
Nunki	51	Sagittarius	σ	18.9	-26	-13	10
Nusakan	202	Corona Borealist	β	15.5	29	56	46
Okab	139	Aquila	ζ	19.1	14	3	47
Paikauhale	117	Scorpius	τ	16.6	-28	13	352
Peacock	43	Pavo	α	20.4	-57	-35	341
Phact	98	Columba	α	5.7	-34	-29	239
Phecda	78	Urda Major	γ	11.6	54	61	141
Pherkad	141	Ursa Minor	γ	15.3	72	41	108
Piautos	329	Cancer	λ	8.3	24	30	199
Pipirima	193	Scorpius	$μ^2$	16.9	-38	4	346
Pleione / 28	310	Taurus	-	3.8	24	-23	167
Polaris	44	Ursa Minor	α	2.5	89	27	123
Polaris Australis	320	Octans	σ	21.1	-89	-28	304
Polis	225	Sagittarius	μ	18.2	-21	-2	10
Pollux	17	Gemini	β	7.8	28	23	192
Porrima	107	Virgo	γ	12.7	-1	61	298
Praecipua / 46	219	Leo Minor	-	10.9	34	64	190
Prima Hyadum	201	Taurus	γ	4.3	16	-24	179
Procyon	8	Canis Minor	α	7.7	5	13	214
Propus	165	Gemini	η	16.2	23	3	189
Ran	212	Eridanus	ε	3.5	-9	-48	196
Rasalas	230	Leo	μ	9.9	26	50	204
Rasalgethi	171	Hercules	$α^1$	17.2	14	28	36
Rasalhague	57	Ophiuchus	α	17.6	13	23	36
Rastaban	113	Draco	β	17.5	52	33	80
Regor[NA]	33	Vela	γ	8.2	-47	-8	263
Regulus	21	Leo	α	10.1	12	49	226
Revati	313	Pisces	ζ	1.2	8	-55	133
Rigel	7	Orion	α	5.2	-8	-25	209
Rigel Kentaurus	3	Centaurus	α	14.7	-60	-1	316
Rotanev	200	Delphinus	β	20.6	15	-16	59
Ruchbah	99	Cassiopeia	δ	1.4	60	-2	127
Rukbat	238	Sagittarius	α	19.4	-41	-23	358
Rutilicus[NA]	133	Hercules	ζ	16.7	32	40	53
Sabik	79	Ophiuchus	η	17.2	-16	14	7
Saclateni	206	Auriga	ζ	5.0	41	-0	165

Table of Coordinates: by Common Name   **198** | P a g e

Common Name	#	Constellation	Gk	RA hrs	Decl. °	Glat. °	Glong. °
Sadachbia	227	Aquarius	γ	22.4	-1	-46	62
Sadalbari	186	Pegasus	μ	22.8	25	-31	91
Sadalmelik	135	Aquarius	α	22.1	-1	-42	60
Sadalsuud	132	Aquarius	β	21.5	-6	-38	48
Sadr	66	Cygnus	γ	20.4	40	2	78
Saiph	56	Orion	κ	5.8	-10	-19	215
Salm	285	Pegasus	τ	23.3	24	-35	98
Sargas	39	Scorpius	θ	17.7	-43	-6	347
Sarin	151	Hercules	δ	17.3	25	31	47
Sceptrum / 53	244	Eridanus	-	4.6	-14	-36	211
Scheat	80	Pegasus	β	23.1	28	-29	96
Schedar	69	Cassiopeia	α	0.7	57	-6	121
Secunda Hyadum	217	Taurus	δ	4.4	18	-22	178
Segin	170	Cassiopeia	ε	1.9	64	2	130
Seginus	145	Bootes	γ	14.5	38	66	67
Sham	271	Sagitta	α	19.6	18	-2	55
Shaula	25	Scorpius	λ	17.6	-37	-2	352
Sheliak	188	Lyra	β	18.8	33	15	63
Sheratan	95	Aries	β	1.9	21	-40	142
Sirius	1	Canis Major	α	6.8	-17	-9	227
Situla	309	Aquarius	κ	22.6	-4	-51	63
Skat	163	Aquarius	δ	22.9	-16	-61	50
Spica	15	Virgo	α	13.4	-11	51	316
Sualocin	215	Delphinus	α	20.7	16	-15	60
Subra	189	Leo	o	9.7	10	42	225
Suhail	67	Vela	λ	9.1	-43	3	266
Sulafat	162	Lyra	γ	19.0	33	13	63
Syrma	248	Virgo	ι	14.3	-6	51	338
Tabit	157	Orion	π^3	4.8	7	-23	191
Taiyangshou	207	Ursa Major	χ	11.8	48	66	150
Taiyi / 8	315	Draco	-	12.9	65	52	122
Talitha	152	Ursa Major	ι	9.0	48	41	172
Tania Australis	149	Ursa Major	μ	10.4	41	56	178
Tania Borealis	179	Ursa Major	λ	10.3	43	55	176
Tarazed	104	Aquila	γ	19.8	11	-7	49
Tarf / Altarf[NA]	191	Cancer	β	8.3	9	23	214

Table of Coordinates: by Common Name   **199** | P a g e

Common Name	#	Constellation	Gk	RA hrs	Decl. °	Glat. °	Glong. °
Taygeta / 19	266	Taurus	q	3.8	24	-24	166
Tegmine	290	Cancer	ζ^1	8.2	18	26	205
Tejat	123	Gemini	μ	6.4	31	4	190
Terebellum	292	Sagittarius	ω	19.9	-26	-25	15
Theemin	220	Eridinus	υ	4.6	-31	-41	231
Thuban	203	Draco	α	14.1	64	51	111
Tiaki	60	Grus	β	22.7	-47	-58	346
Tianguan	136	Taurus	ζ	5.6	21	-6	186
Tianyi / 7	318	Draco	-	12.8	67	50	124
Titawin	251	Andromeda	υ	1.6	41	-21	132
Tonatiuh	328	Camelopardalis	-	12.1	77	40	126
Torcular	265	Pisces	o	1.8	9	-51	145
Tsih[NA]	82	Cassiopeia	γ	0.9	61	-2	124
Tureis	119	Puppis	ρ	8.1	-24	5	243
Ukdah	233	Hydra	ι	9.7	-1	36	236
Unukalhai	94	Serpens	α	15.7	6	44	14
Unurgunite	184	Canis Major	σ	7.0	-28	-10	239
Vega	5	Lyra	α	18.6	38	19	67
Veritate / 14	314	Andromeda	-	23.5	39	-21	106
Vindemaitrix	122	Virgo	ε	13.0	11	74	312
Wasat	185	Gemini	δ	7.3	22	16	196
Wazn	153	Columba	β	5.8	-36	-27	241
Wezen	36	Canis Major	δ	7.1	-26	-8	238
Wurren	245	Phoenix	ζ	1.1	-55	-62	298
Xamidimura	142	Scorpius	$μ^1$	16.9	-38	4	346
Xuange	259	Bootes	λ	14.3	46	65	87
Yed Posterior	160	Ophiuchus	ε	16.3	-4.7	31	9
Yed Prior	106	Ophiuchus	δ	16.2	-3.7	32	9
Yildun	270	Ursa Minor	δ	17.5	87	28	119
Zaniah	232	Virgo	η	12.3	-1	61	286
Zaurak	137	Eridanus	γ	4.0	-13	-45	205
Zavijava	197	Virgo	β	11.8	46	61	271
Zhang	254	Hydra	$υ^1$	9.9	-15	29	251
Zibal	298	Eridanus	ζ	3.3	-9	-51	192
Zosma	88	Leo	δ	11.2	21	67	224
Zubenelgenubi	109	Libra	$α^2$	14.8	-16	38	340
Zubenelhakradi	234	Libra	γ	15.6	-15	32	352

Table of Coordinates: by Common Name   **200** | P a g e

Common Name	#	Constellation	Gk	RA hrs	Decl. °	Glat. °	Glong. °
Zubene-schamali	93	Libra	β	15.3	-9	39	352

Table of Coordinates: by Constellation

Common Name	#	Constellation	Gk	RA hrs	Decl. °	Glat. °	Glong. °
Adhil	303	Andromeda	ξ	1.4	46	-17	129
Almach	59	Andromeda	γ	2.1	42	-19	137
Alpheratz	53	Andromeda	α	0.1	29	-33	112
Mirach	55	Andromeda	β	1.2	36	-27	127
Nembus / 51 / υ Persesu	196	Andromeda	-	1.6	49	-14	131
Titawin	251	Andromeda	υ	1.6	41	-21	132
Veritate / 14	314	Andromeda	-	23.5	39	-21	106
Albali	216	Aquarius	ε	20.8	-9	-30	38
Ancha	257	Aquarius	θ	22.3	47	-49	54
Bunda	297	Aquarius	ξ	21.6	-8	-40	46
Sadachbia	227	Aquarius	γ	22.4	-1	-46	62
Sadalmelik	135	Aquarius	α	22.1	-1	-42	60
Sadalsuud	132	Aquarius	β	21.5	-6	-38	48
Situla	309	Aquarius	κ	22.6	-4	-51	63
Skat	163	Aquarius	δ	22.9	-16	-61	50
Alshain	208	Aquila	β	19.9	6	-11	46
Altair	12	Aquila	α	19.8	9	-9	48
Deneb El Okab[NA]	245	Aquila	ε	19.0	15	5	47
Libertas	293	Aquila	ξ	19.9	8	-10	48
Okab	139	Aquila	ζ	19.1	14	3	47
Tarazed	104	Aquila	γ	19.8	11	-7	49
55	>333	Aquila	η	19.9	1	-13	41
65	161	Aquila	θ	20.2	-1	-18	42
-	119	Ara	β	17.4	-55	-11	335
-	119	Ara	α	17.5	-50	-9	341
Cervantes	312	Ara	μ	17.7	-52	-11	341
Bharani / 41	198	Aries	-	2.8	27	-29	153
Botein	269	Aries	δ	3.2	20	-32	163
Hamal	49	Aries	α	2.1	23	-36	145
Lilii Borea / 39	282	Aries	-	2.8	29	-27	151

Common Name	#	Constellation	Gk	RA hrs	Decl. °	Glat. °	Glong. °
Mesarthim	294	Aries	γ^1	1.9	19	-41	143
Sheratan	95	Aries	β	1.9	21	-40	142
37	>333	Aries	o	2.7	15	-39	159
Almaaz	144	Auriga	ϵ	5.0	44	1.2	163
Capella	6	Auriga	α	5.3	46	5	163
Haedus	155	Auriga	η	5.1	41	0	165
Hassaleh	101	Auriga	ι	5.0	33	-6	171
Mahasim	97	Auriga	θ	6.0	37	7	174
Menkalinan	40	Auriga	β	6.0	45	10	167
Saclateni	206	Auriga	ζ	5.0	41	-0	165
Alkalurops	267	Bootes	μ^2	15.4	37	56	60
Arcturus	4	Bootes	α	14.3	19	69	15
Izar	75	Bootes	ϵ	14.7	27	65	39
Merga / 38	327	Bootes	-	14.8	46	60	80
Muphrid	100	Bootes	η	13.9	18	5	73
Nekkar	183	Bootes	β	15.0	40	60	68
Seginus	145	Bootes	γ	14.5	38	66	67
Xuange	259	Bootes	λ	14.3	46	65	87
30	>333	Bootes	ζ	14.7	14	61	11
49	180	Bootes	δ	15.3	33	58	53
Tonatiuh	328	Camelopardalis	-	12.1	77	40	126
Acubens	262	Cancer	α	9.0	12	34	217
Asellus Asutralis	236	Cancer	δ	8.7	18	33	208
Asellus Borealis	288	Cancer	γ	8.7	21	34	204
Copernicus / 55	330	Cancer	-	8.9	28	38	197
Meleph	331	Cancer	ϵ	8.7	20	32	206
Nahn	324	Cancer	ξ	9.2	22	40	206
Piautos	329	Cancer	λ	8.3	24	30	199
Tarf / Altarf[NA]	191	Cancer	β	8.3	9	23	214
Tegmine	290	Cancer	ζ^1	8.2	18	26	205
Chara	260	Canes Venatici	β	12.6	41	75	136
Cor Caroli	127	Canes Venatici	α^2	12.9	38	79	118
La Superba	317	Canes Venatici	γ	12.8	45	72	127
Adhara	23	Canis Major	ϵ	7.0	-29	-11	240
Aludra	82	Canis Major	η	7.6	-29	-7	243
Furud	143	Canis Major	ζ	6.3	-30	-19	238
Mirzam	46	Canis Major	β	6.4	-18	-14	226

Table of Coordinates: by Constellation   **203** | P a g e

Common Name	#	Constellation	Gk	RA hrs	Decl. °	Glat. °	Glong. °
Muliphein	253	Canis Major	γ	7.1	-16	-4	228
Sirius	1	Canis Major	α	6.8	-17	-9	227
Unurgunite	184	Canis Major	σ	7.0	-28	-10	239
Wezen	36	Canis Major	δ	7.1	-26	-8	238
Gomeisa	129	Canis Minor	β	7.5	8	12	209
Procyon	8	Canis Minor	α	7.7	5	13	214
Algedi	194	Capricornus	α²	20.3	-13	-25	31
Alshat	296	Capricornus	ν	20.3	-13	-25	31
Dabih	147	Capricornus	β¹	20.4	-15	-26	29
Deneb Algedi	121	Capricornus	δ	21.8	-16	-46	38
Nashira	205	Capricornus	γ	21.7	-17	-45	36
-	107	Carina	θ	10.7	-64	-5	290
-	133	Carina	υ	9.8	-65	-9	285
Aspidiske / Turais[NA]	62	Carina	ι	9.3	-59	-7	279
Avior	38	Carina	ε	8.4	-59	-13	274
Canopus	2	Carina	α	6.4	-53	-25	261
Miaplacidus	28	Carina	β	9.2	-70	-14	286
Achird	180	Cassiopeia	η	0.8	58	-5	123
Caph	71	Cassiopeia	β	0.2	59	-3	117
Castula	287	Cassiopeia	υ²	0.9	59	-4	124
Fulu	204	Cassiopeia	ζ	0.6	54	-9	121
Ruchbah	99	Cassiopeia	δ	1.4	60	-2	127
Schedar	69	Cassiopeia	α	0.7	57	-6	121
Segin	170	Cassiopeia	ε	1.9	64	2	130
Tsih[NA]	82	Cassiopeia	γ	0.9	61	-2	124
-	74	Centaurus	ε	13.7	-53	9	310
-	78	Centaurus	η	14.6	-42	17	323
-	91	Centaurus	δ	12.1	-51	12	296
-	110	Centaurus	ι	13.3	-37	26	309
Hadar	11	Centaurus	β	14.1	-60	1	312
Menkent	52	Centaurus	θ	14.1	-36	24	319
Muhlifain[NA]	62	Centaurus	γ	12.7	-49	14	301
Rigel Kentaurus	3	Centaurus	α	14.7	-60	-1	316
Alderamin	81	Cepheus	α	21.3	63	9	101
Alfirk	159	Cepheus	β	21.5	71	14	108
Errai	158	Cepheus	γ	23.7	78	15	119

Common Name	#	Constellation	Gk	RA hrs	Decl. °	Glat. °	Glong. °
Kurhah	263	Cepheus	ξ	22.1	38	7	106
Baten Kaitos	213	Ceteus	ζ	1.9	-10	-68	166
Kaffaljidhma	192	Ceteus	γ	2.7	14	-49	169
Mira	332	Ceteus	o	2.3	-3	-58	168
Diphda	50	Cetus	β	0.7	-18	-81	111
Menkar	86	Cetus	α	3.0	4.1	-46	173
Elkurud	306	Columba	θ	6.1	-37	-24	244
Phact	98	Columba	α	5.7	-34	-29	239
Wazn	153	Columba	β	5.8	-36	-27	241
Diadem	301	Coma Berenices	α	3.2	18	79	328
HIP 62752 A	>333	Coma Berenices	-	12.9	27	90	120
Meridiana	252	Corona Australis	α	19.2	-38	-20	359
Alphecca / Gemma[NA]	64	Corona Borealis	α	15.6	27	54	42
Nusakan	202	Corona Borealist	β	15.5	29	56	46
Alchiba	243	Corvus	α	12.1	-25	37	291
Algorab	134	Corvus	δ	12.5	-17	46	295
Gienah / Gienah Ghurab[NA]	90	Corvus	γ	12.3	-17	45	291
Kraz	96	Corvus	β	12.6	-23	39	298
Alkes	249	Crater	α	11	-18	37	269
Acrux	13	Crux	α	12.4	-63	-0	300
Gacrux	24	Crux	γ	12.5	-57	6	300
Ginan	195	Crux	ε	12.4	-60	2	299
Imai	108	Crux	δ	12.3	-59	4	298
Mimosa	20	Crux	β	12.8	-60	33	303
Albireo	146	Cygnus	β¹	19.5	28	5	62
Aljanah / Gienah Cygni[NA]	84	Cygnus	ε	20.8	34	-6	76
Azelfafage	291	Cygnus	π¹	21.7	51	-1	95
Deneb	19	Cygnus	α	20.7	45	2.0	84
Fawaris	131	Cygnus	δ	19.7	45	10	79
Sadr	66	Cygnus	γ	20.4	40	2	78

Table of Coordinates: by Constellation   **205** | P a g e

Common Name	#	Constellation	Gk	RA hrs	Decl. °	Glat. °	Glong. °
63	>333	Cygnus	-	21.1	48	0	89
Gal. 0, 90		Cygnus	-	21.2	48	0	90
Aldulfin	246	Delphinus	ε	20.6	11	-17	55
Musica / 18	321	Delphinus	-	21.0	11	-22	59
Rotanev	200	Delphinus	β	20.6	15	-16	59
Sualocin	215	Delphinus	α	20.7	16	-15	60
Aldhibah	154	Draco	ζ	17.1	66	35	96
Alrakis	323	Draco	μ	17.1	54	37	82
Alruba	325	Draco	-	17.7	54	31	81
Alsafi	289	Draco	σ	19.5	70	22	101
Altais	150	Draco	δ	19.2	40	23	99
Athebyne / Alhibain[NA]	105	Draco	η	16.4	61	41	93
Dziban	284	Draco	ψ¹	17.7	72	31	103
Edasich	164	Draco	ι	15.4	59	49	94
Eltanin	68	Draco	γ	17.9	51	29	79
Fafnir / 42	299	Draco	-	18.4	66	27	95
Giausar	222	Draco	λ	11.5	69	+46	133
Grumium	211	Draco	ξ	17.9	57	30	85
Rastaban	113	Draco	β	17.5	52	33	80
Taiyi / 8	315	Draco	-	12.9	65	52	122
Thuban	203	Draco	α	14.1	64	51	111
Tianyi / 7	318	Draco	-	12.8	67	50	124
Kitalpha	235	Equulus	α	21.3	5	-29	56
Acamar	124	Eridanus	θ¹	3.0	-40	-61	248
Achernar	10	Eridanus	α	1.6	-57	-59	291
Angetenar	295	Eridanus	τ²	2.9	-21	-62	207
Azha	231	Eridanus	η	2.9	-8.9	-55	187
Beemim	239	Eridanus	υ³	4.4	-34	-44	235
Beid	247	Eridanus	σ¹	4.2	-7	-38	199
Cursa / Kursa[NA]	111	Eridanus	β	5.1	-5	-25	205
Keid	276	Eridanus	o	4.3	-8	-38	201
Ran	212	Eridanus	ε	3.5	-9	-48	196
Sceptrum / 53	244	Eridanus	-	4.6	-14	-36	211
Zaurak	137	Eridanus	γ	4.0	-13	-45	205
Zibal	298	Eridanus	ζ	3.3	-9	-51	192
Theemin	220	Eridinus	υ	4.6	-31	-41	231

Common Name	#	Constellation	Gk	RA hrs	Decl. °	Glat. °	Glong. °
Dalim / 12 Eridanus	226	Fornax	α	3.2	-29	-59	225
Alhena / Almeisan[NA]	42	Gemini	γ	6.6	16	5	197
Alzirr	168	Gemini	ξ	6.8	13	5	201
Castor	45	Gemini	α	7.6	32	23	187
Jishui	304	Gemini	o	7.7	35	24	185
Mebsuta	148	Gemini	ε	6.7	25	10	189
Mekbuda	242	Gemini	ζ	7.1	21	12	196
Pollux	17	Gemini	β	7.8	28	23	192
Propus	165	Gemini	η	16.2	23	3	189
Tejat	123	Gemini	μ	6.4	31	4	190
Wasat	185	Gemini	δ	7.3	22	16	196
Aldhanab	140	Grus	γ	21.9	-37	-51	6
Alnair	30	Grus	α	22.1	-47	-53	350
Tiaki	60	Grus	β	22.7	-47	-58	346
Cujam	283	Hercules	ω	16.4	14	39	29
Kornephoros	112	Hercules	β	16.5	21	40	39
Maasym	273	Hercules	λ	17.5	26	28	49
Marsic	307	Hercules	κ	16.1	17	44	31
Rasalgethi	171	Hercules	α¹	17.2	14	28	36
Rutilicus[NA]	133	Hercules	ζ	16.7	32	40	53
Sarin	151	Hercules	δ	17.3	25	31	47
67	154	Hercules	π	17.3	37	34	61
85	>333	Hercules	ι	17.7	46	31	72
-	117	Hydra	β	0.4	-77	-40	305
Alphard	47	Hydra	α	9.5	-9	29	241
Ashlesha	183	Hydra	ε	8.8	6	29	221
Felis	305	Hydra	-	9.9	-19	27	255
Minchir	278	Hydra	σ	8.6	3	25	223
Ukdah	233	Hydra	ι	9.7	-1	36	236
Zhang	254	Hydra	υ¹	9.9	-15	29	251
-	122	Hydrus	α	2.0	-62	54	289
Adhafera	177	Leo	ζ	10.3	23	55	210
Algieba	92	Leo	γ¹	10.3	20	55	217
Alterf	269	Leo	λ	9.5	23	45	207
Chertan	167	Leo	θ	11.2	15.4	65	235

Common Name	#	Constellation	Gk	RA hrs	Decl. °	Glat. °	Glong. °
Denebola	61	Leo	β	11.8	15	7`	251
Rasalas	230	Leo	μ	9.9	26	50	204
Regulus	21	Leo	α	10.1	12	49	226
Subra	189	Leo	o	9.7	10	42	225
Zosma	88	Leo	δ	11.2	21	67	224
17	137	Leo	ε	9.8	24	48	207
Praecipua / 46	219	Leo Minor	-	10.9	34	64	190
Arneb	89	Lepus	α	5.5	-18	-25	221
Nihal	115	Lepus	β	5.5	-21	-27	224
Brachium	161	Libra	σ	15.1	-25	29	337
Zubenelgenubi	109	Libra	α²	14.8	-16	38	340
Zubenelhakradi	234	Libra	γ	15.6	-15	32	352
Zubene-schamali	93	Libra	β	15.3	-9	39	352
24	>333	Libra	ι	15.2	-20	32	343
-	114	Lupus	γ	15.6	-41	12	333
KakkabNA	75	Lupus	α	14.7	-43	11	322
KekouanNA	111	Lupus	β	15.0	-43	14	326
Alsciaukat / 31	261	Lynx	-	8.4	43	34	177
38	>333	Lynx	-	9.3	37	45	187
40	153	Lynx	α	9.4	34	45	190
Aladfar	275	Lyra	η	19.2	39	13	71
Sheliak	188	Lyra	β	18.8	33	15	63
Sulafat	162	Lyra	γ	19.0	33	13	63
Vega	5	Lyra	α	18.6	38	19	67
-	101	Musca	α	12.6	-69	-6	301
Polaris Australis	320	Octans	σ	21.1	-89	-28	304
Marfik	223	Opheuchus	λ	16.5	2	32	17
Cebalrai	110	Ophiuchus	β	17.6	5	17	29
Guniibuu / 36	311	Ophiuchus	-	17.3	-27	7	358
Rasalhague	57	Ophiuchus	α	17.6	13	23	36
Sabik	79	Ophiuchus	η	17.2	-16	14	7
Yed Posterior	160	Ophiuchus	ε	16.3	-4.7	31	9
Yed Prior	106	Ophiuchus	δ	16.2	-3.7	32	9
13	86	Ophiuchus	ζ	16.6	-11	24	6
Alnilam	29	Orion	ε	5.6	-1	-17	205
Alnitak	31	Orion	ζ	5.7	-2	-17	207

Table of Coordinates: by Constellation   **208** | P a g e

Common Name	#	Constellation	Gk	RA hrs	Decl. °	Glat. °	Glong. °
Bellatrix	26	Orion	γ	5.4	6	-16	197
Betelgeuse	9	Orion	β	5.9	7	-9	200
Hatysa	114	Orion	ι	5.6	-6	-20	210
Meissa / Heka[NA]	173	Orion	λ	5.6	10	-12	195
Mintaka	70	Orion	δ	5.5	-0	-18	204
Rigel	7	Orion	α	5.2	-8	-25	209
Saiph	56	Orion	κ	5.8	-10	-19	215
Tabit	157	Orion	π³	4.8	7	-23	191
Peacock	43	Pavo	α	20.4	-57	-35	341
Algenib	118	Pegasus	γ	0.2	15	-47	109
Alkarab	274	Pegasus	υ	23.4	23	-35	99
Biham	187	Pegasus	θ	22.2	6	-39	67
Enif	76	Pegasus	ε	21.7	10	-31	66
Helvetios / 51	322	Pegasus	-	23.0	21	-35	90
Homam	175	Pegasus	ζ	22.7	11	-41	79
Markab	85	Pegasus	α	23.1	15	-40	88
Matar	133	Pegasus	η	22.7	30	-25	93
Sadalbari	186	Pegasus	μ	22.8	25	-31	91
Salm	285	Pegasus	τ	23.3	24	-35	98
Scheat	80	Pegasus	β	23.1	28	-29	96
Algol	58	Perseus	β	3.1	41	-15	3
Atik	224	Perseus	o	3.7	32	-18	160
Menkib	240	Perseus	ξ	4.0	36	-13	160
Miram	214	Perseus	η	2.8	56	-3	138
Mirfak	34	Perseus	α	3.4	50	-6	147
Misam	218	Perseus	κ	3.2	45	-11	147
23	133	Perseus	γ	3.1	53	-4	142
44	119	Perseus	ζ	3.9	32	-17	162
45	133	Perseus	ε	4.0	40	-10	157
Ankaa	77	Phoenix	α	0.4	-42	-74	320
Wurren	245	Phoenix	ζ	1.1	-55	-62	298
Alpherg	126	Pisces	η	1.5	45	-46	137
Alrescha	221	Pisces	α	2.0	3	-56	155
Fumalsamakah	279	Pisces	β	23.1	4	-50	79
Revati	313	Pisces	ζ	1.2	8	-55	133
Torcular	265	Pisces	o	1.8	9	-51	145
Fomalhaut	18	Piscis Austrinus	α	23.0	-30	-65	21

Table of Coordinates: by Constellation

Common Name	#	Constellation	Gk	RA hrs	Decl. °	Glat. °	Glong. °
-	102	Puppis	π	7.3	-37	-11	249
Altaleban[NA]	157	Puppis	τ	6.8	-51	-21	260
Azmidi	178	Puppis	ξ	7.8	-25	1	241
Naos	63	Puppis	ζ	8.1	-40	-5	256
Tureis	119	Puppis	ρ	8.1	-24	5	243
Sham	271	Sagitta	α	19.6	18	-2	55
Ainalrami	302	Sagittarius	ν	18.9	-23	-11	13
Albaldah	125	Sagittarius	π	19.2	-21	-13	16
Alnasl	138	Sagittarius	γ^2	18.1	-30	-5	1
Arkab Posterior	264	Sagittarius	β^2	19.4	-45	-24	353
Arkab Prior	237	Sagittarius	β^1	19.4	-44	-24	354
Ascella	91	Sagittarius	ζ	19.0	-30	-15	7
Kaus Australis	33	Sagittarius	ε	18.4	-34	-10	359
Kaus Borealis	116	Sagittarius	λ	16.5	-25.4	-6.5	7.7
Kaus Media	103	Sagittarius	δ	18.3	-29.8	-7.2	3.0
Nunki	51	Sagittarius	σ	18.9	-26	-13	10
Polis	225	Sagittarius	μ	18.2	-21	-2	10
Rukbat	238	Sagittarius	α	19.4	-41	-23	358
Terebellum	292	Sagittarius	ω	19.9	-26	-25	15
Gal. 0, 0		Sagittarius	-	17.8	-29	0	0
Acrab / Graffias[NA]	87	Scorpius	β^1	16.1	-20	24	353
Alniyat	130	Scorpius	σ	16.4	-26	17	351
Antares	16	Scorpius	α	16.5	-26	15	-8
Dschubba	72	Scorpius	δ	16.0	-23	23	350
Fang	128	Scorpius	π	16.0	-26	20	347
Fuyue	156	Scorpius	η	5.1	14	0	165
Girtab[NA]	81	Scorpius	κ	17.6	-39	-5	351
Iklil	228	Scorpius	ρ	15.9	-29	18	345
Jabbah	280	Scorpius	ν	16.2	-19	23	355
Larawag	73	Scorpius	ε	16.8	-34	7	349
Lesath	102	Scorpius	υ	17.5	-37	-2	351
Paikauhale	117	Scorpius	τ	16.6	-28	13	352
Pipirima	193	Scorpius	μ^2	16.9	-38	4	346
Sargas	39	Scorpius	θ	17.7	-43	-6	347
Shaula	25	Scorpius	λ	17.6	-37	-2	352
Xamidimura	142	Scorpius	μ^1	16.9	-38	4	346

Common Name	#	Constellation	Gk	RA hrs	Decl. °	Glat. °	Glong. °
HIP 4363	>333	Sculptor	-	0.9	-27	-89	114
Alya	286	Serpens	θ¹	18.9	4.2	1	37
Gudja	250	Serpens	κ	15.6	18	48	30
Unukalhai	94	Serpens	α	15.7	6	44	14
Ain	190	Taurus	ε	4.5	19	-20	178
Alcyone	120	Taurus	η	3.8	24	-23	167
Aldebaran	14	Taurus	α	4.6	17	-20	181
Asterope / 21	326	Taurus	-	3.8	25	-23	166
Atlas / 27	199	Taurus	-	3.8	24	-23	167
Celaeno / 16	319	Taurus	-	3.7	24	-24	166
Chamukuy	210	Taurus	θ	4.5	16	-22	180
Electra / 17	209	Taurus	-	3.7	24	-24	166
Elnath	27	Taurus	β	5.4	29	-4	178
Maia / 20	229	Taurus	-	3.8	24	-24	166
Merope / 23	255	Taurus	-	3.8	24	-24	167
Pleione / 28	310	Taurus	-	3.8	24	-23	167
Prima Hyadum	201	Taurus	γ	4.3	16	-24	179
Secunda Hyadum	217	Taurus	δ	4.4	18	-22	178
Taygeta / 19	266	Taurus	q	3.8	24	-24	166
Tianguan	136	Taurus	ζ	5.6	21	-6	186
94	>333	Taurus	τ	4.7	23	-15	177
Gal. 0, 180		Taurus	-	5.8	29	0	180
Mothallah	176	Triangulum	α	1.9	30	-31	139
4	142	Triangulum	β	2.2	35	-25	141
-	121	Triangulum Australe	β	15.9	-63	-7	322
-	123	Triangulum Australe	γ	15.3	-69	-9	316
Atria	41	Triangulum Australe	α	16.8	-69	-15	321
-	123	Tucana	α	22.3	-61	-48	330
Phecda	78	Urda Major	γ	11.6	54	61	141
Alcor	241	Ursa Major	γ	13.6	55	61	113
Alioth	32	Ursa Major	ε	12.9	56	61	122
Alkaid / Benetnasch[NA]	37	Ursa Major	η	13.8	49	65	101

Table of Coordinates: by Constellation   **211** | P a g e

Common Name	#	Constellation	Gk	RA hrs	Decl. °	Glat. °	Glong. °
Alkaphrah	256	Ursa Major	κ	9.1	47	42	173
Alula Australis	272	Ursa Major	ξ	11.3	32	69	195
Alula Borealis	182	Ursa Major	ν	11.3	33	69	191
Chalawan / 47	308	Ursa Major	-	11.0	40	63	176
Dubhe	35	Ursa Major	α	11.1	62	51	143
Intercrus / 41 Lynx^NA	316	Ursa Major	-	9.5	46	46	174
Megrez	166	Ursa Major	δ	12.3	57	59	133
Merak	74	Ursa Major	β	11.0	56	55	149
Mizar	65	Ursa Major	ζ	13.4	55	62	113
Muscida	169	Ursa Major	o	8.5	61	35	156
Taiyangshou	207	Ursa Major	χ	11.8	48	66	150
Talitha	152	Ursa Major	ι	9.0	48	41	172
Tania Australis	149	Ursa Major	μ	10.4	41	56	178
Tania Borealis	179	Ursa Major	λ	10.3	43	55	176
Kochab	54	Ursa Minor	β	14.8	74	74	113
Pherkad	141	Ursa Minor	γ	15.3	72	41	108
Polaris	44	Ursa Minor	α	2.5	89	27	123
Yildun	270	Ursa Minor	δ	17.5	87	28	119
-	102	Vela	μ	11	-49	9	283
Alsephina	48	Vela	δ	8.7	-55	-7	272
Markeb	83	Vela	κ	9.4	-55	-3	276
Regor^NA	33	Vela	γ	8.2	-47	-8	263
Suhail	67	Vela	λ	9.1	-43	3	266
Gal. 0, 270		Vela	-	9.2	-48	0	270
Elgafar	300	Virgo	φ	14.5	-2	53	345
Heze	172	Virgo	ζ	13.6	-1	60	325
Kang	258	Virgo	κ	14.2	-10	48	334
Khambalia	281	Virgo	λ	14.3	-13	44	333
Minelauva	174	Virgo	δ	12.9	3	66	306
Porrima	107	Virgo	γ	12.7	-1	61	298
Spica	15	Virgo	α	13.4	-11	51	316
Syrma	248	Virgo	ι	14.3	-6	51	338
Vindemaitrix	122	Virgo	ε	13.0	11	74	312
Zaniah	232	Virgo	η	12.3	-1	61	286
Zavijava	197	Virgo	β	11.8	46	61	271
107	>333	Virgo	μ	14.7	-6	47	347

Table of Coordinates: by Constellation   **212** | P a g e

Common Name	#	Constellation	Gk	RA hrs	Decl. °	Glat. °	Glong. °
109	>333	Virgo	-	14.8	2	53	355
Anser	277	Vulcan	α	19.5	25	3	59

Table of Coordinates: by *RA-Decl*

Common Name	#	Constellation	Gk	RA hrs	Decl. °	Glat. °	Glong. °
Alpheratz	53	Andromeda	α	0.1	29	-33	112
Algenib	118	Pegasus	γ	0.2	15	-47	109
Caph	71	Cassiopeia	β	0.2	59	-3	117
-	117	Hydra	β	0.4	-77	-40	305
Ankaa	77	Phoenix	α	0.4	-42	-74	320
Fulu	204	Cassiopeia	ζ	0.6	54	-9	121
Diphda	50	Cetus	β	0.7	-18	-81	111
Schedar	69	Cassiopeia	α	0.7	57	-6	121
Achird	180	Cassiopeia	η	0.8	58	-5	123
HIP 4363	>333	Sculptor	-	0.9	-27	-89	114
Castula	287	Cassiopeia	υ²	0.9	59	-4	124
Tsih^NA	82	Cassiopeia	γ	0.9	61	-2	124
Wurren	245	Phoenix	ζ	1.1	-55	-62	298
Revati	313	Pisces	ζ	1.2	8	-55	133
Mirach	55	Andromeda	β	1.2	36	-27	127
Adhil	303	Andromeda	ξ	1.4	46	-17	129
Ruchbah	99	Cassiopeia	δ	1.4	60	-2	127
Alpherg	126	Pisces	η	1.5	45	-46	137
Achernar	10	Eridanus	α	1.6	-57	-59	291
Titawin	251	Andromeda	υ	1.6	41	-21	132
Nembus / 51 / υ Persesu	196	Andromeda	-	1.6	49	-14	131
Torcular	265	Pisces	o	1.8	9	-51	145
Baten Kaitos	213	Ceteus	ζ	1.9	-10	-68	166
Mesarthim	294	Aries	γ¹	1.9	19	-41	143
Sheratan	95	Aries	β	1.9	21	-40	142
Mothallah	176	Triangulum	α	1.9	30	-31	139
Segin	170	Cassiopeia	ε	1.9	64	2	130
-	122	Hydrus	α	2.0	-62	54	289
Alrescha	221	Pisces	α	2.0	3	-56	155

Common Name	#	Constellation	Gk	RA hrs	Decl. °	Glat. °	Glong. °
Hamal	49	Aries	α	2.1	23	-36	145
Almach	59	Andromeda	γ	2.1	42	-19	137
4	142	Triangulum	β	2.2	35	-25	141
Mira	332	Ceteus	o	2.3	-3	-58	168
Polaris	44	Ursa Minor	α	2.5	89	27	123
Kaffaljidhma	192	Ceteus	γ	2.7	14	-49	169
37	>333	Aries	o	2.7	15	-39	159
Bharani / 41	198	Aries	-	2.8	27	-29	153
Lilii Borea / 39	282	Aries	-	2.8	29	-27	151
Miram	214	Perseus	η	2.8	56	-3	138
Angetenar	295	Eridanus	τ²	2.9	-21	-62	207
Azha	231	Eridanus	η	2.9	-8.9	-55	187
Acamar	124	Eridanus	θ¹	3.0	-40	-61	248
Menkar	86	Cetus	α	3.0	4.1	-46	173
Algol	58	Perseus	β	3.1	41	-15	3
23	133	Perseus	γ	3.1	53	-4	142
Dalim / 12 Eridanus	226	Fornax	α	3.2	-29	-59	225
Diadem	301	Coma Berenices	α	3.2	18	79	328
Botein	269	Aries	δ	3.2	20	-32	163
Misam	218	Perseus	κ	3.2	45	-11	147
Zibal	298	Eridanus	ζ	3.3	-9	-51	192
Mirfak	34	Perseus	α	3.4	50	-6	147
Ran	212	Eridanus	ε	3.5	-9	-48	196
Celaeno / 16	319	Taurus	-	3.7	24	-24	166
Electra / 17	209	Taurus	-	3.7	24	-24	166
Atik	224	Perseus	o	3.7	32	-18	160
Alcyone	120	Taurus	η	3.8	24	-23	167
Atlas / 27	199	Taurus	-	3.8	24	-23	167
Maia / 20	229	Taurus	-	3.8	24	-24	166
Merope / 23	255	Taurus	-	3.8	24	-24	167
Pleione / 28	310	Taurus	-	3.8	24	-23	167
Taygeta / 19	266	Taurus	q	3.8	24	-24	166
Asterope / 21	326	Taurus	-	3.8	25	-23	166
44	119	Perseus	ζ	3.9	32	-17	162
Zaurak	137	Eridanus	γ	4.0	-13	-45	205

Table of Coordinates: by RA-Decl

Common Name	#	Constellation	Gk	RA hrs	Decl. °	Glat. °	Glong. °
Menkib	240	Perseus	ξ	4.0	36	-13	160
45	133	Perseus	ε	4.0	40	-10	157
Beid	247	Eridanus	σ¹	4.2	-7	-38	199
Keid	276	Eridanus	o	4.3	-8	-38	201
Prima Hyadum	201	Taurus	γ	4.3	16	-24	179
Beemim	239	Eridanus	υ³	4.4	-34	-44	235
Secunda Hyadum	217	Taurus	δ	4.4	18	-22	178
Chamukuy	210	Taurus	θ	4.5	16	-22	180
Ain	190	Taurus	ε	4.5	19	-20	178
Theemin	220	Eridinus	υ	4.6	-31	-41	231
Sceptrum / 53	244	Eridanus	-	4.6	-14	-36	211
Aldebaran	14	Taurus	α	4.6	17	-20	181
94	>333	Taurus	τ	4.7	23	-15	177
Tabit	157	Orion	π³	4.8	7	-23	191
Hassaleh	101	Auriga	ι	5.0	33	-6	171
Saclateni	206	Auriga	ζ	5.0	41	-0	165
Almaaz	144	Auriga	ε	5.0	44	1.2	163
Cursa / Kursa[NA]	111	Eridanus	β	5.1	-5	-25	205
Fuyue	156	Scorpius	η	5.1	14	0	165
Haedus	155	Auriga	η	5.1	41	0	165
Rigel	7	Orion	α	5.2	-8	-25	209
Capella	6	Auriga	α	5.3	46	5	163
Bellatrix	26	Orion	γ	5.4	6	-16	197
Elnath	27	Taurus	β	5.4	29	-4	178
Nihal	115	Lepus	β	5.5	-21	-27	224
Arneb	89	Lepus	α	5.5	-18	-25	221
Mintaka	70	Orion	δ	5.5	-0	-18	204
Hatysa	114	Orion	ι	5.6	-6	-20	210
Alnilam	29	Orion	ε	5.6	-1	-17	205
Meissa / Heka[NA]	173	Orion	λ	5.6	10	-12	195
Tianguan	136	Taurus	ζ	5.6	21	-6	186
Phact	98	Columba	α	5.7	-34	-29	239
Alnitak	31	Orion	ζ	5.7	-2	-17	207
Wazn	153	Columba	β	5.8	-36	-27	241
Saiph	56	Orion	κ	5.8	-10	-19	215
Gal. 0, 180		Taurus	-	5.8	29	0	180

Common Name	#	Constellation	Gk	RA hrs	Decl. °	Glat. °	Glong. °
Betelgeuse	9	Orion	β	5.9	7	-9	200
Mahasim	97	Auriga	θ	6.0	37	7	174
Menkalinan	40	Auriga	β	6.0	45	10	167
Elkurud	306	Columba	θ	6.1	-37	-24	244
Furud	143	Canis Major	ζ	6.3	-30	-19	238
Canopus	2	Carina	α	6.4	-53	-25	261
Mirzam	46	Canis Major	β	6.4	-18	-14	226
Tejat	123	Gemini	μ	6.4	31	4	190
Alhena / Almeisan[NA]	42	Gemini	γ	6.6	16	5	197
Mebsuta	148	Gemini	ε	6.7	25	10	189
Altaleban[NA]	157	Puppis	τ	6.8	-51	-21	260
Sirius	1	Canis Major	α	6.8	-17	-9	227
Alzirr	168	Gemini	ξ	6.8	13	5	201
Adhara	23	Canis Major	ε	7.0	-29	-11	240
Unurgunite	184	Canis Major	σ	7.0	-28	-10	239
Wezen	36	Canis Major	δ	7.1	-26	-8	238
Muliphein	253	Canis Major	γ	7.1	-16	-4	228
Mekbuda	242	Gemini	ζ	7.1	21	12	196
-	102	Puppis	π	7.3	-37	-11	249
Wasat	185	Gemini	δ	7.3	22	16	196
Gomeisa	129	Canis Minor	β	7.5	8	12	209
Aludra	82	Canis Major	η	7.6	-29	-7	243
Castor	45	Gemini	α	7.6	32	23	187
Procyon	8	Canis Minor	α	7.7	5	13	214
Jishui	304	Gemini	o	7.7	35	24	185
Azmidi	178	Puppis	ξ	7.8	-25	1	241
Pollux	17	Gemini	β	7.8	28	23	192
Naos	63	Puppis	ζ	8.1	-40	-5	256
Tureis	119	Puppis	ρ	8.1	-24	5	243
Regor[NA]	33	Vela	γ	8.2	-47	-8	263
Tegmine	290	Cancer	ζ¹	8.2	18	26	205
Tarf / Altarf[NA]	191	Cancer	β	8.3	9	23	214
Piautos	329	Cancer	λ	8.3	24	30	199
Avior	38	Carina	ε	8.4	-59	-13	274
Alsciaukat / 31	261	Lynx	-	8.4	43	34	177
Muscida	169	Ursa Major	o	8.5	61	35	156

Common Name	#	Constellation	Gk	RA hrs	Decl. °	Glat. °	Glong. °
Minchir	278	Hydra	σ	8.6	3	25	223
Alsephina	48	Vela	δ	8.7	-55	-7	272
Asellus Asutralis	236	Cancer	δ	8.7	18	33	208
Meleph	331	Cancer	ε	8.7	20	32	206
Asellus Borealis	288	Cancer	γ	8.7	21	34	204
Ashlesha	183	Hydra	ε	8.8	6	29	221
Copernicus / 55	330	Cancer	-	8.9	28	38	197
Acubens	262	Cancer	α	9.0	12	34	217
Talitha	152	Ursa Major	ι	9.0	48	41	172
Suhail	67	Vela	λ	9.1	-43	3	266
Alkaphrah	256	Ursa Major	κ	9.1	47	42	173
Miaplacidus	28	Carina	β	9.2	-70	-14	286
Gal. 0, 270		Vela	-	9.2	-48	0	270
Nahn	324	Cancer	ξ	9.2	22	40	206
Aspidiske / Turais[NA]	62	Carina	ι	9.3	-59	-7	279
38	>333	Lynx	-	9.3	37	45	187
Markeb	83	Vela	κ	9.4	-55	-3	276
40	153	Lynx	α	9.4	34	45	190
Alphard	47	Hydra	α	9.5	-9	29	241
Alterf	269	Leo	λ	9.5	23	45	207
Intercrus / 41 Lynx[NA]	316	Ursa Major	-	9.5	46	46	174
Ukdah	233	Hydra	ι	9.7	-1	36	236
Subra	189	Leo	o	9.7	10	42	225
-	133	Carina	υ	9.8	-65	-9	285
17	137	Leo	ε	9.8	24	48	207
Felis	305	Hydra	-	9.9	-19	27	255
Zhang	254	Hydra	υ¹	9.9	-15	29	251
Rasalas	230	Leo	μ	9.9	26	50	204
Regulus	21	Leo	α	10.1	12	49	226
Algieba	92	Leo	γ¹	10.3	20	55	217
Adhafera	177	Leo	ζ	10.3	23	55	210
Tania Borealis	179	Ursa Major	λ	10.3	43	55	176
Tania Australis	149	Ursa Major	μ	10.4	41	56	178
-	107	Carina	θ	10.7	-64	-5	290
Praecipua / 46	219	Leo Minor	-	10.9	34	64	190

Common Name	#	Constellation	Gk	RA hrs	Decl. °	Glat. °	Glong. °
-	102	Vela	μ	11	-49	9	283
Alkes	249	Crater	α	11	-18	37	269
Chalawan / 47	308	Ursa Major	-	11.0	40	63	176
Merak	74	Ursa Major	β	11.0	56	55	149
Dubhe	35	Ursa Major	α	11.1	62	51	143
Chertan	167	Leo	θ	11.2	15.4	65	235
Zosma	88	Leo	δ	11.2	21	67	224
Alula Australis	272	Ursa Major	ξ	11.3	32	69	195
Alula Borealis	182	Ursa Major	ν	11.3	33	69	191
Giausar	222	Draco	λ	11.5	69	+46	133
Phecda	78	Urda Major	γ	11.6	54	61	141
Denebola	61	Leo	β	11.8	15	7`	251
Zavijava	197	Virgo	β	11.8	46	61	271
Taiyangshou	207	Ursa Major	χ	11.8	48	66	150
-	91	Centaurus	δ	12.1	-51	12	296
Alchiba	243	Corvus	α	12.1	-25	37	291
Tonatiuh	328	Camelopardalis	-	12.1	77	40	126
Imai	108	Crux	δ	12.3	-59	4	298
Gienah / Gienah Ghurab[NA]	90	Corvus	γ	12.3	-17	45	291
Zaniah	232	Virgo	η	12.3	-1	61	286
Megrez	166	Ursa Major	δ	12.3	57	59	133
Acrux	13	Crux	α	12.4	-63	-0	300
Ginan	195	Crux	ε	12.4	-60	2	299
Gacrux	24	Crux	γ	12.5	-57	6	300
Algorab	134	Corvus	δ	12.5	-17	46	295
-	101	Musca	α	12.6	-69	-6	301
Kraz	96	Corvus	β	12.6	-23	39	298
Chara	260	Canes Venatici	β	12.6	41	75	136
Muhlifain[NA]	62	Centaurus	γ	12.7	-49	14	301
Porrima	107	Virgo	γ	12.7	-1	61	298
Mimosa	20	Crux	β	12.8	-60	33	303
La Superba	317	Canes Venatici	γ	12.8	45	72	127
Tianyi / 7	318	Draco	-	12.8	67	50	124
Minelauva	174	Virgo	δ	12.9	3	66	306
HIP 62752 A	>333	Coma Berenices	-	12.9	27	90	120

Table of Coordinates: by RA-Decl

Common Name	#	Constellation	Gk	RA hrs	Decl. °	Glat. °	Glong. °
Cor Caroli	127	Canes Venatici	α^2	12.9	38	79	118
Alioth	32	Ursa Major	ε	12.9	56	61	122
Taiyi / 8	315	Draco	-	12.9	65	52	122
Vindemaitrix	122	Virgo	ε	13.0	11	74	312
-	110	Centaurus	ι	13.3	-37	26	309
Spica	15	Virgo	α	13.4	-11	51	316
Mizar	65	Ursa Major	ζ	13.4	55	62	113
Heze	172	Virgo	ζ	13.6	-1	60	325
Alcor	241	Ursa Major	γ	13.6	55	61	113
-	74	Centaurus	ε	13.7	-53	9	310
Alkaid / Benetnasch[NA]	37	Ursa Major	η	13.8	49	65	101
Muphrid	100	Bootes	η	13.9	18	5	73
Hadar	11	Centaurus	β	14.1	-60	1	312
Menkent	52	Centaurus	θ	14.1	-36	24	319
Thuban	203	Draco	α	14.1	64	51	111
Kang	258	Virgo	κ	14.2	-10	48	334
Khambalia	281	Virgo	λ	14.3	-13	44	333
Syrma	248	Virgo	ι	14.3	-6	51	338
Arcturus	4	Bootes	α	14.3	19	69	15
Xuange	259	Bootes	λ	14.3	46	65	87
Elgafar	300	Virgo	φ	14.5	-2	53	345
Seginus	145	Bootes	γ	14.5	38	66	67
-	78	Centaurus	η	14.6	-42	17	323
Rigel Kentaurus	3	Centaurus	α	14.7	-60	-1	316
Kakkab[NA]	75	Lupus	α	14.7	-43	11	322
107	>333	Virgo	μ	14.7	-6	47	347
30	>333	Bootes	ζ	14.7	14	61	11
Izar	75	Bootes	ε	14.7	27	65	39
Zubenelgenubi	109	Libra	α^2	14.8	-16	38	340
109	>333	Virgo	-	14.8	2	53	355
Merga / 38	327	Bootes	-	14.8	46	60	80
Kochab	54	Ursa Minor	β	14.8	74	74	113
Kekouan[NA]	111	Lupus	β	15.0	-43	14	326
Nekkar	183	Bootes	β	15.0	40	60	68
Brachium	161	Libra	σ	15.1	-25	29	337
24	>333	Libra	ι	15.2	-20	32	343

Table of Coordinates: by RA-Decl

Common Name	#	Constellation	Gk	RA hrs	Decl. °	Glat. °	Glong. °
-	123	Triangulum Australe	γ	15.3	-69	-9	316
Zubene-schamali	93	Libra	β	15.3	-9	39	352
49	180	Bootes	δ	15.3	33	58	53
Pherkad	141	Ursa Minor	γ	15.3	72	41	108
Alkalurops	267	Bootes	μ²	15.4	37	56	60
Edasich	164	Draco	ι	15.4	59	49	94
Nusakan	202	Corona Borealist	β	15.5	29	56	46
-	114	Lupus	γ	15.6	-41	12	333
Zubenelhakradi	234	Libra	γ	15.6	-15	32	352
Gudja	250	Serpens	κ	15.6	18	48	30
Alphecca / Gemma[NA]	64	Corona Borealis	α	15.6	27	54	42
Unukalhai	94	Serpens	α	15.7	6	44	14
-	121	Triangulum Australe	β	15.9	-63	-7	322
Iklil	228	Scorpius	ρ	15.9	-29	18	345
Fang	128	Scorpius	π	16.0	-26	20	347
Dschubba	72	Scorpius	δ	16.0	-23	23	350
Acrab / Graffias[NA]	87	Scorpius	β¹	16.1	-20	24	353
Marsic	307	Hercules	κ	16.1	17	44	31
Jabbah	280	Scorpius	ν	16.2	-19	23	355
Yed Prior	106	Ophiuchus	δ	16.2	-3.7	32	9
Propus	165	Gemini	η	16.2	23	3	189
Yed Posterior	160	Ophiuchus	ε	16.3	-4.7	31	9
Alniyat	130	Scorpius	σ	16.4	-26	17	351
Cujam	283	Hercules	ω	16.4	14	39	29
Athebyne / Alhibain[NA]	105	Draco	η	16.4	61	41	93
Antares	16	Scorpius	α	16.5	-26	15	-8
Kaus Borealis	116	Sagittarius	λ	16.5	-25.4	-6.5	7.7
Marfik	223	Opheuchus	λ	16.5	2	32	17
Kornephoros	112	Hercules	β	16.5	21	40	39
Paikauhale	117	Scorpius	τ	16.6	-28	13	352

Common Name	#	Constellation	Gk	RA hrs	Decl. °	Glat. °	Glong. °
13	86	Ophiuchus	ζ	16.6	-11	24	6
Rutilicus^{NA}	133	Hercules	ζ	16.7	32	40	53
Atria	41	Triangulum Australe	α	16.8	-69	-15	321
Larawag	73	Scorpius	ε	16.8	-34	7	349
Pipirima	193	Scorpius	μ²	16.9	-38	4	346
Xamidimura	142	Scorpius	μ¹	16.9	-38	4	346
Alrakis	323	Draco	μ	17.1	54	37	82
Aldhibah	154	Draco	ζ	17.1	66	35	96
Sabik	79	Ophiuchus	η	17.2	-16	14	7
Rasalgethi	171	Hercules	α¹	17.2	14	28	36
Guniibuu / 36	311	Ophiuchus	-	17.3	-27	7	358
Sarin	151	Hercules	δ	17.3	25	31	47
67	154	Hercules	π	17.3	37	34	61
-	119	Ara	β	17.4	-55	-11	335
-	119	Ara	α	17.5	-50	-9	341
Lesath	102	Scorpius	υ	17.5	-37	-2	351
Maasym	273	Hercules	λ	17.5	26	28	49
Rastaban	113	Draco	β	17.5	52	33	80
Yildun	270	Ursa Minor	δ	17.5	87	28	119
Girtab^{NA}	81	Scorpius	κ	17.6	-39	-5	351
Shaula	25	Scorpius	λ	17.6	-37	-2	352
Cebalrai	110	Ophiuchus	β	17.6	5	17	29
Rasalhague	57	Ophiuchus	α	17.6	13	23	36
Cervantes	312	Ara	μ	17.7	-52	-11	341
Sargas	39	Scorpius	θ	17.7	-43	-6	347
85	>333	Hercules	ι	17.7	46	31	72
Alruba	325	Draco	-	17.7	54	31	81
Dziban	284	Draco	ψ¹	17.7	72	31	103
Gal. 0, 0		Sagittarius	-	17.8	-29	0	0
Eltanin	68	Draco	γ	17.9	51	29	79
Grumium	211	Draco	ξ	17.9	57	30	85
Alnasl	138	Sagittarius	γ²	18.1	-30	-5	1
Polis	225	Sagittarius	μ	18.2	-21	-2	10
Kaus Media	103	Sagittarius	δ	18.3	-29.8	-7.2	3.0
Kaus Australis	33	Sagittarius	ε	18.4	-34	-10	359
Fafnir / 42	299	Draco	-	18.4	66	27	95

Table of Coordinates: by RA-Decl   **222** | P a g e

Common Name	#	Constellation	Gk	RA hrs	Decl. °	Glat. °	Glong. °
Vega	5	Lyra	α	18.6	38	19	67
Sheliak	188	Lyra	β	18.8	33	15	63
Nunki	51	Sagittarius	σ	18.9	-26	-13	10
Ainalrami	302	Sagittarius	ν	18.9	-23	-11	13
Alya	286	Serpens	θ¹	18.9	4.2	1	37
Ascella	91	Sagittarius	ζ	19.0	-30	-15	7
Deneb El Okab^{NA}	245	Aquila	ε	19.0	15	5	47
Sulafat	162	Lyra	γ	19.0	33	13	63
Okab	139	Aquila	ζ	19.1	14	3	47
Meridiana	252	Corona Australis	α	19.2	-38	-20	359
Albaldah	125	Sagittarius	π	19.2	-21	-13	16
Aladfar	275	Lyra	η	19.2	39	13	71
Altais	150	Draco	δ	19.2	40	23	99
Arkab Posterior	264	Sagittarius	β²	19.4	-45	-24	353
Arkab Prior	237	Sagittarius	β¹	19.4	-44	-24	354
Rukbat	238	Sagittarius	α	19.4	-41	-23	358
Anser	277	Vulcan	α	19.5	25	3	59
Albireo	146	Cygnus	β¹	19.5	28	5	62
Alsafi	289	Draco	σ	19.5	70	22	101
Sham	271	Sagitta	α	19.6	18	-2	55
Fawaris	131	Cygnus	δ	19.7	45	10	79
Altair	12	Aquila	α	19.8	9	-9	48
Tarazed	104	Aquila	γ	19.8	11	-7	49
Terebellum	292	Sagittarius	ω	19.9	-26	-25	15
55	>333	Aquila	η	19.9	1	-13	41
Alshain	208	Aquila	β	19.9	6	-11	46
Libertas	293	Aquila	ξ	19.9	8	-10	48
65	161	Aquila	θ	20.2	-1	-18	42
Algedi	194	Capricornus	α²	20.3	-13	-25	31
Alshat	296	Capricornus	ν	20.3	-13	-25	31
Peacock	43	Pavo	α	20.4	-57	-35	341
Dabih	147	Capricornus	β¹	20.4	-15	-26	29
Sadr	66	Cygnus	γ	20.4	40	2	78
Aldulfin	246	Delphinus	ε	20.6	11	-17	55
Rotanev	200	Delphinus	β	20.6	15	-16	59

Table of Coordinates: by RA-Decl

Common Name	#	Constellation	Gk	RA hrs	Decl. °	Glat. °	Glong. °
Sualocin	215	Delphinus	α	20.7	16	-15	60
Deneb	19	Cygnus	α	20.7	45	2.0	84
Albali	216	Aquarius	ε	20.8	-9	-30	38
Aljanah / Gienah Cygni[NA]	84	Cygnus	ε	20.8	34	-6	76
Musica / 18	321	Delphinus	-	21.0	11	-22	59
Polaris Australis	320	Octans	σ	21.1	-89	-28	304
63	>333	Cygnus	-	21.1	48	0	89
Gal. 0, 90		Cygnus	-	21.2	48	0	90
Kitalpha	235	Equulus	α	21.3	5	-29	56
Alderamin	81	Cepheus	α	21.3	63	9	101
Sadalsuud	132	Aquarius	β	21.5	-6	-38	48
Alfirk	159	Cepheus	β	21.5	71	14	108
Bunda	297	Aquarius	ξ	21.6	-8	-40	46
Nashira	205	Capricornus	γ	21.7	-17	-45	36
Enif	76	Pegasus	ε	21.7	10	-31	66
Azelfafage	291	Cygnus	π¹	21.7	51	-1	95
Deneb Algedi	121	Capricornus	δ	21.8	-16	-46	38
Aldhanab	140	Grus	γ	21.9	-37	-51	6
Alnair	30	Grus	α	22.1	-47	-53	350
Sadalmelik	135	Aquarius	α	22.1	-1	-42	60
Kurhah	263	Cepheus	ξ	22.1	38	7	106
Biham	187	Pegasus	θ	22.2	6	-39	67
-	123	Tucana	α	22.3	-61	-48	330
Ancha	257	Aquarius	θ	22.3	47	-49	54
Sadachbia	227	Aquarius	γ	22.4	-1	-46	62
Situla	309	Aquarius	κ	22.6	-4	-51	63
Tiaki	60	Grus	β	22.7	-47	-58	346
Homam	175	Pegasus	ζ	22.7	11	-41	79
Matar	133	Pegasus	η	22.7	30	-25	93
Sadalbari	186	Pegasus	μ	22.8	25	-31	91
Skat	163	Aquarius	δ	22.9	-16	-61	50
Fomalhaut	18	Piscis Austrinus	α	23.0	-30	-65	21
Helvetios / 51	322	Pegasus	-	23.0	21	-35	90
Fumalsamakah	279	Pisces	β	23.1	4	-50	79
Markab	85	Pegasus	α	23.1	15	-40	88
Scheat	80	Pegasus	β	23.1	28	-29	96

Table of Coordinates: by RA-Decl

Common Name	#	Constellation	Gk	RA hrs	Decl. °	Glat. °	Glong. °
Salm	285	Pegasus	τ	23.3	24	-35	98
Alkarab	274	Pegasus	υ	23.4	23	-35	99
Veritate / 14	314	Andromeda	-	23.5	39	-21	106
Errai	158	Cepheus	γ	23.7	78	15	119

Table of Coordinates: by *Decl-RA*

Common Name	#	Constellation	Gk	RA hrs	Decl. °	Glat. °	Glong. °
Polaris Australis	320	Octans	σ	21.1	-89	-28	304
-	117	Hydra	β	0.4	-77	-40	305
Miaplacidus	28	Carina	β	9.2	-70	-14	286
-	101	Musca	α	12.6	-69	-6	301
-	123	Triangulum Australe	γ	15.3	-69	-9	316
Atria	41	Triangulum Australe	α	16.8	-69	-15	321
-	133	Carina	υ	9.8	-65	-9	285
-	107	Carina	θ	10.7	-64	-5	290
Acrux	13	Crux	α	12.4	-63	-0	300
-	121	Triangulum Australe	β	15.9	-63	-7	322
-	122	Hydrus	α	2.0	-62	54	289
-	123	Tucana	α	22.3	-61	-48	330
Ginan	195	Crux	ε	12.4	-60	2	299
Mimosa	20	Crux	β	12.8	-60	33	303
Hadar	11	Centaurus	β	14.1	-60	1	312
Rigel Kentaurus	3	Centaurus	α	14.7	-60	-1	316
Avior	38	Carina	ε	8.4	-59	-13	274
Aspidiske / Turais[NA]	62	Carina	ι	9.3	-59	-7	279
Imai	108	Crux	δ	12.3	-59	4	298
Achernar	10	Eridanus	α	1.6	-57	-59	291
Gacrux	24	Crux	γ	12.5	-57	6	300
Peacock	43	Pavo	α	20.4	-57	-35	341
Wurren	245	Phoenix	ζ	1.1	-55	-62	298
Alsephina	48	Vela	δ	8.7	-55	-7	272
Markeb	83	Vela	κ	9.4	-55	-3	276
-	119	Ara	β	17.4	-55	-11	335
Canopus	2	Carina	α	6.4	-53	-25	261
-	74	Centaurus	ε	13.7	-53	9	310

Common Name	#	Constellation	Gk	RA hrs	Decl. °	Glat. °	Glong. °
Cervantes	312	Ara	μ	17.7	-52	-11	341
Altaleban[NA]	157	Puppis	τ	6.8	-51	-21	260
-	91	Centaurus	δ	12.1	-51	12	296
-	119	Ara	α	17.5	-50	-9	341
-	102	Vela	μ	11	-49	9	283
Muhlifain[NA]	62	Centaurus	γ	12.7	-49	14	301
Gal. 0, 270		Vela	-	9.2	-48	0	270
Regor[NA]	33	Vela	γ	8.2	-47	-8	263
Alnair	30	Grus	α	22.1	-47	-53	350
Tiaki	60	Grus	β	22.7	-47	-58	346
Arkab Posterior	264	Sagittarius	β²	19.4	-45	-24	353
Arkab Prior	237	Sagittarius	β¹	19.4	-44	-24	354
Suhail	67	Vela	λ	9.1	-43	3	266
Kakkab[NA]	75	Lupus	α	14.7	-43	11	322
Kekouan[NA]	111	Lupus	β	15.0	-43	14	326
Sargas	39	Scorpius	θ	17.7	-43	-6	347
Ankaa	77	Phoenix	α	0.4	-42	-74	320
-	78	Centaurus	η	14.6	-42	17	323
-	114	Lupus	γ	15.6	-41	12	333
Rukbat	238	Sagittarius	α	19.4	-41	-23	358
Acamar	124	Eridanus	θ¹	3.0	-40	-61	248
Naos	63	Puppis	ζ	8.1	-40	-5	256
Girtab[NA]	81	Scorpius	κ	17.6	-39	-5	351
Pipirima	193	Scorpius	μ²	16.9	-38	4	346
Xamidimura	142	Scorpius	μ¹	16.9	-38	4	346
Meridiana	252	Corona Australis	α	19.2	-38	-20	359
Elkurud	306	Columba	θ	6.1	-37	-24	244
-	102	Puppis	π	7.3	-37	-11	249
-	110	Centaurus	ι	13.3	-37	26	309
Lesath	102	Scorpius	υ	17.5	-37	-2	351
Shaula	25	Scorpius	λ	17.6	-37	-2	352
Aldhanab	140	Grus	γ	21.9	-37	-51	6
Wazn	153	Columba	β	5.8	-36	-27	241
Menkent	52	Centaurus	θ	14.1	-36	24	319
Beemim	239	Eridanus	υ³	4.4	-34	-44	235
Phact	98	Columba	α	5.7	-34	-29	239

Common Name	#	Constellation	Gk	RA hrs	Decl. °	Glat. °	Glong. °
Larawag	73	Scorpius	ε	16.8	-34	7	349
Kaus Australis	33	Sagittarius	ε	18.4	-34	-10	359
Theemin	220	Eridinus	υ	4.6	-31	-41	231
Furud	143	Canis Major	ζ	6.3	-30	-19	238
Alnasl	138	Sagittarius	γ²	18.1	-30	-5	1
Ascella	91	Sagittarius	ζ	19.0	-30	-15	7
Fomalhaut	18	Piscis Austrinus	α	23.0	-30	-65	21
Kaus Media	103	Sagittarius	δ	18.3	-29.8	-7.2	3.0
Dalim / 12 Eridanus	226	Fornax	α	3.2	-29	-59	225
Adhara	23	Canis Major	ε	7.0	-29	-11	240
Aludra	82	Canis Major	η	7.6	-29	-7	243
Iklil	228	Scorpius	ρ	15.9	-29	18	345
Gal. 0, 0		Sagittarius	-	17.8	-29	0	0
Unurgunite	184	Canis Major	σ	7.0	-28	-10	239
Paikauhale	117	Scorpius	τ	16.6	-28	13	352
HIP 4363	>333	Sculptor	-	0.9	-27	-89	114
Guniibuu / 36	311	Ophiuchus	-	17.3	-27	7	358
Wezen	36	Canis Major	δ	7.1	-26	-8	238
Fang	128	Scorpius	π	16.0	-26	20	347
Alniyat	130	Scorpius	σ	16.4	-26	17	351
Antares	16	Scorpius	α	16.5	-26	15	-8
Nunki	51	Sagittarius	σ	18.9	-26	-13	10
Terebellum	292	Sagittarius	ω	19.9	-26	-25	15
Kaus Borealis	116	Sagittarius	λ	16.5	-25.4	-6.5	7.7
Azmidi	178	Puppis	ξ	7.8	-25	1	241
Alchiba	243	Corvus	α	12.1	-25	37	291
Brachium	161	Libra	σ	15.1	-25	29	337
Tureis	119	Puppis	ρ	8.1	-24	5	243
Kraz	96	Corvus	β	12.6	-23	39	298
Dschubba	72	Scorpius	δ	16.0	-23	23	350
Ainalrami	302	Sagittarius	ν	18.9	-23	-11	13
Angetenar	295	Eridanus	τ²	2.9	-21	-62	207
Nihal	115	Lepus	β	5.5	-21	-27	224
Polis	225	Sagittarius	μ	18.2	-21	-2	10
Albaldah	125	Sagittarius	π	19.2	-21	-13	16
24	>333	Libra	ι	15.2	-20	32	343

Table of Coordinates: by Decl-RA

Common Name	#	Constellation	Gk	RA hrs	Decl. °	Glat. °	Glong. °
Acrab / GraffiasNA	87	Scorpius	β^1	16.1	-20	24	353
Felis	305	Hydra	-	9.9	-19	27	255
Jabbah	280	Scorpius	ν	16.2	-19	23	355
Diphda	50	Cetus	β	0.7	-18	-81	111
Arneb	89	Lepus	α	5.5	-18	-25	221
Mirzam	46	Canis Major	β	6.4	-18	-14	226
Alkes	249	Crater	α	11	-18	37	269
Sirius	1	Canis Major	α	6.8	-17	-9	227
Gienah / Gienah GhurabNA	90	Corvus	γ	12.3	-17	45	291
Algorab	134	Corvus	δ	12.5	-17	46	295
Nashira	205	Capricornus	γ	21.7	-17	-45	36
Muliphein	253	Canis Major	γ	7.1	-16	-4	228
Zubenelgenubi	109	Libra	α^2	14.8	-16	38	340
Sabik	79	Ophiuchus	η	17.2	-16	14	7
Deneb Algedi	121	Capricornus	δ	21.8	-16	-46	38
Skat	163	Aquarius	δ	22.9	-16	-61	50
Zhang	254	Hydra	υ^1	9.9	-15	29	251
Zubenelhakradi	234	Libra	γ	15.6	-15	32	352
Dabih	147	Capricornus	β^1	20.4	-15	-26	29
Sceptrum / 53	244	Eridanus	-	4.6	-14	-36	211
Zaurak	137	Eridanus	γ	4.0	-13	-45	205
Khambalia	281	Virgo	λ	14.3	-13	44	333
Algedi	194	Capricornus	α^2	20.3	-13	-25	31
Alshat	296	Capricornus	ν	20.3	-13	-25	31
Spica	15	Virgo	α	13.4	-11	51	316
13	86	Ophiuchus	ζ	16.6	-11	24	6
Baten Kaitos	213	Ceteus	ζ	1.9	-10	-68	166
Saiph	56	Orion	κ	5.8	-10	-19	215
Kang	258	Virgo	κ	14.2	-10	48	334
Zibal	298	Eridanus	ζ	3.3	-9	-51	192
Ran	212	Eridanus	ε	3.5	-9	-48	196
Alphard	47	Hydra	α	9.5	-9	29	241
Zubene-schamali	93	Libra	β	15.3	-9	39	352
Albali	216	Aquarius	ε	20.8	-9	-30	38

Common Name	#	Constellation	Gk	RA hrs	Decl. °	Glat. °	Glong. °
Azha	231	Eridanus	η	2.9	-8.9	-55	187
Keid	276	Eridanus	o	4.3	-8	-38	201
Rigel	7	Orion	α	5.2	-8	-25	209
Bunda	297	Aquarius	ξ	21.6	-8	-40	46
Beid	247	Eridanus	σ¹	4.2	-7	-38	199
Hatysa	114	Orion	ι	5.6	-6	-20	210
Syrma	248	Virgo	ι	14.3	-6	51	338
107	>333	Virgo	μ	14.7	-6	47	347
Sadalsuud	132	Aquarius	β	21.5	-6	-38	48
Cursa / Kursa^NA	111	Eridanus	β	5.1	-5	-25	205
Yed Posterior	160	Ophiuchus	ε	16.3	-4.7	31	9
Situla	309	Aquarius	κ	22.6	-4	-51	63
Yed Prior	106	Ophiuchus	δ	16.2	-3.7	32	9
Mira	332	Ceteus	o	2.3	-3	-58	168
Alnitak	31	Orion	ζ	5.7	-2	-17	207
Elgafar	300	Virgo	φ	14.5	-2	53	345
Alnilam	29	Orion	ε	5.6	-1	-17	205
Ukdah	233	Hydra	ι	9.7	-1	36	236
Zaniah	232	Virgo	η	12.3	-1	61	286
Porrima	107	Virgo	γ	12.7	-1	61	298
Heze	172	Virgo	ζ	13.6	-1	60	325
65	161	Aquila	θ	20.2	-1	-18	42
Sadalmelik	135	Aquarius	α	22.1	-1	-42	60
Sadachbia	227	Aquarius	γ	22.4	-1	-46	62
Mintaka	70	Orion	δ	5.5	-0	-18	204
55	>333	Aquila	η	19.9	1	-13	41
109	>333	Virgo	-	14.8	2	53	355
Marfik	223	Opheuchus	λ	16.5	2	32	17
Alrescha	221	Pisces	α	2.0	3	-56	155
Minchir	278	Hydra	σ	8.6	3	25	223
Minelauva	174	Virgo	δ	12.9	3	66	306
Fumalsamakah	279	Pisces	β	23.1	4	-50	79
Menkar	86	Cetus	α	3.0	4.1	-46	173
Alya	286	Serpens	θ¹	18.9	4.2	1	37
Procyon	8	Canis Minor	α	7.7	5	13	214
Cebalrai	110	Ophiuchus	β	17.6	5	17	29
Kitalpha	235	Equulus	α	21.3	5	-29	56

Table of Coordinates: by Decl-RA

Common Name	#	Constellation	Gk	RA hrs	Decl. °	Glat. °	Glong. °
Bellatrix	26	Orion	γ	5.4	6	-16	197
Ashlesha	183	Hydra	ε	8.8	6	29	221
Unukalhai	94	Serpens	α	15.7	6	44	14
Alshain	208	Aquila	β	19.9	6	-11	46
Biham	187	Pegasus	θ	22.2	6	-39	67
Tabit	157	Orion	π^3	4.8	7	-23	191
Betelgeuse	9	Orion	β	5.9	7	-9	200
Revati	313	Pisces	ζ	1.2	8	-55	133
Gomeisa	129	Canis Minor	β	7.5	8	12	209
Libertas	293	Aquila	ξ	19.9	8	-10	48
Torcular	265	Pisces	o	1.8	9	-51	145
Tarf / Altarf[NA]	191	Cancer	β	8.3	9	23	214
Altair	12	Aquila	α	19.8	9	-9	48
Meissa / Heka[NA]	173	Orion	λ	5.6	10	-12	195
Subra	189	Leo	o	9.7	10	42	225
Enif	76	Pegasus	ε	21.7	10	-31	66
Vindemaitrix	122	Virgo	ε	13.0	11	74	312
Tarazed	104	Aquila	γ	19.8	11	-7	49
Aldulfin	246	Delphinus	ε	20.6	11	-17	55
Musica / 18	321	Delphinus	-	21.0	11	-22	59
Homam	175	Pegasus	ζ	22.7	11	-41	79
Acubens	262	Cancer	α	9.0	12	34	217
Regulus	21	Leo	α	10.1	12	49	226
Alzirr	168	Gemini	ξ	6.8	13	5	201
Rasalhague	57	Ophiuchus	α	17.6	13	23	36
Kaffaljidhma	192	Ceteus	γ	2.7	14	-49	169
Fuyue	156	Scorpius	η	5.1	14	0	165
30	>333	Bootes	ζ	14.7	14	61	11
Cujam	283	Hercules	ω	16.4	14	39	29
Rasalgethi	171	Hercules	α^1	17.2	14	28	36
Okab	139	Aquila	ζ	19.1	14	3	47
Algenib	118	Pegasus	γ	0.2	15	-47	109
37	>333	Aries	o	2.7	15	-39	159
Denebola	61	Leo	β	11.8	15	7`	251
Deneb El Okab[NA]	245	Aquila	ε	19.0	15	5	47
Rotanev	200	Delphinus	β	20.6	15	-16	59

Table of Coordinates: by Decl-RA

Common Name	#	Constellation	Gk	RA hrs	Decl. °	Glat. °	Glong. °
Markab	85	Pegasus	α	23.1	15	-40	88
Chertan	167	Leo	θ	11.2	15.4	65	235
Prima Hyadum	201	Taurus	γ	4.3	16	-24	179
Chamukuy	210	Taurus	θ	4.5	16	-22	180
Alhena / Almeisan^{NA}	42	Gemini	γ	6.6	16	5	197
Sualocin	215	Delphinus	α	20.7	16	-15	60
Aldebaran	14	Taurus	α	4.6	17	-20	181
Marsic	307	Hercules	κ	16.1	17	44	31
Diadem	301	Coma Berenices	α	3.2	18	79	328
Secunda Hyadum	217	Taurus	δ	4.4	18	-22	178
Tegmine	290	Cancer	ζ¹	8.2	18	26	205
Asellus Asutralis	236	Cancer	δ	8.7	18	33	208
Muphrid	100	Bootes	η	13.9	18	5	73
Gudja	250	Serpens	κ	15.6	18	48	30
Sham	271	Sagitta	α	19.6	18	-2	55
Mesarthim	294	Aries	γ¹	1.9	19	-41	143
Ain	190	Taurus	ε	4.5	19	-20	178
Arcturus	4	Bootes	α	14.3	19	69	15
Botein	269	Aries	δ	3.2	20	-32	163
Meleph	331	Cancer	ε	8.7	20	32	206
Algieba	92	Leo	γ¹	10.3	20	55	217
Sheratan	95	Aries	β	1.9	21	-40	142
Tianguan	136	Taurus	ζ	5.6	21	-6	186
Mekbuda	242	Gemini	ζ	7.1	21	12	196
Asellus Borealis	288	Cancer	γ	8.7	21	34	204
Zosma	88	Leo	δ	11.2	21	67	224
Kornephoros	112	Hercules	β	16.5	21	40	39
Helvetios / 51	322	Pegasus	-	23.0	21	-35	90
Wasat	185	Gemini	δ	7.3	22	16	196
Nahn	324	Cancer	ξ	9.2	22	40	206
Hamal	49	Aries	α	2.1	23	-36	145
94	>333	Taurus	τ	4.7	23	-15	177
Alterf	269	Leo	λ	9.5	23	45	207
Adhafera	177	Leo	ζ	10.3	23	55	210

Common Name	#	Constellation	Gk	RA hrs	Decl. °	Glat. °	Glong. °
Propus	165	Gemini	η	16.2	23	3	189
Alkarab	274	Pegasus	υ	23.4	23	-35	99
Celaeno / 16	319	Taurus	-	3.7	24	-24	166
Electra / 17	209	Taurus	-	3.7	24	-24	166
Alcyone	120	Taurus	η	3.8	24	-23	167
Atlas / 27	199	Taurus	-	3.8	24	-23	167
Maia / 20	229	Taurus	-	3.8	24	-24	166
Merope / 23	255	Taurus	-	3.8	24	-24	167
Pleione / 28	310	Taurus	-	3.8	24	-23	167
Taygeta / 19	266	Taurus	q	3.8	24	-24	166
Piautos	329	Cancer	λ	8.3	24	30	199
17	137	Leo	ε	9.8	24	48	207
Salm	285	Pegasus	τ	23.3	24	-35	98
Asterope / 21	326	Taurus	-	3.8	25	-23	166
Mebsuta	148	Gemini	ε	6.7	25	10	189
Sarin	151	Hercules	δ	17.3	25	31	47
Anser	277	Vulcan	α	19.5	25	3	59
Sadalbari	186	Pegasus	μ	22.8	25	-31	91
Rasalas	230	Leo	μ	9.9	26	50	204
Maasym	273	Hercules	λ	17.5	26	28	49
Bharani / 41	198	Aries	-	2.8	27	-29	153
HIP 62752 A	>333	Coma Berenices	-	12.9	27	90	120
Izar	75	Bootes	ε	14.7	27	65	39
Alphecca / Gemma^NA	64	Corona Borealis	α	15.6	27	54	42
Pollux	17	Gemini	β	7.8	28	23	192
Copernicus / 55	330	Cancer	-	8.9	28	38	197
Albireo	146	Cygnus	β1	19.5	28	5	62
Scheat	80	Pegasus	β	23.1	28	-29	96
Alpheratz	53	Andromeda	α	0.1	29	-33	112
Lilii Borea / 39	282	Aries	-	2.8	29	-27	151
Elnath	27	Taurus	β	5.4	29	-4	178
Gal. 0, 180		Taurus	-	5.8	29	0	180
Nusakan	202	Corona Borealist	β	15.5	29	56	46
Mothallah	176	Triangulum	α	1.9	30	-31	139

Table of Coordinates: by Decl-RA   **233** | P a g e

Common Name	#	Constellation	Gk	RA hrs	Decl. °	Glat. °	Glong. °
Matar	133	Pegasus	η	22.7	30	-25	93
Tejat	123	Gemini	μ	6.4	31	4	190
Atik	224	Perseus	o	3.7	32	-18	160
44	119	Perseus	ζ	3.9	32	-17	162
Castor	45	Gemini	α	7.6	32	23	187
Alula Australis	272	Ursa Major	ξ	11.3	32	69	195
RutilicusNA	133	Hercules	ζ	16.7	32	40	53
Hassaleh	101	Auriga	ι	5.0	33	-6	171
Alula Borealis	182	Ursa Major	ν	11.3	33	69	191
49	180	Bootes	δ	15.3	33	58	53
Sheliak	188	Lyra	β	18.8	33	15	63
Sulafat	162	Lyra	γ	19.0	33	13	63
40	153	Lynx	α	9.4	34	45	190
Praecipua / 46	219	Leo Minor	-	10.9	34	64	190
Aljanah / Gienah CygniNA	84	Cygnus	ε	20.8	34	-6	76
4	142	Triangulum	β	2.2	35	-25	141
Jishui	304	Gemini	o	7.7	35	24	185
Mirach	55	Andromeda	β	1.2	36	-27	127
Menkib	240	Perseus	ξ	4.0	36	-13	160
Mahasim	97	Auriga	θ	6.0	37	7	174
38	>333	Lynx	-	9.3	37	45	187
Alkalurops	267	Bootes	μ²	15.4	37	56	60
67	154	Hercules	π	17.3	37	34	61
Cor Caroli	127	Canes Venatici	α²	12.9	38	79	118
Seginus	145	Bootes	γ	14.5	38	66	67
Vega	5	Lyra	α	18.6	38	19	67
Kurhah	263	Cepheus	ξ	22.1	38	7	106
Aladfar	275	Lyra	η	19.2	39	13	71
Veritate / 14	314	Andromeda	-	23.5	39	-21	106
45	133	Perseus	ε	4.0	40	-10	157
Chalawan / 47	308	Ursa Major	-	11.0	40	63	176
Nekkar	183	Bootes	β	15.0	40	60	68
Altais	150	Draco	δ	19.2	40	23	99
Sadr	66	Cygnus	γ	20.4	40	2	78
Titawin	251	Andromeda	υ	1.6	41	-21	132
Algol	58	Perseus	β	3.1	41	-15	3

Common Name	#	Constellation	Gk	RA hrs	Decl. °	Glat. °	Glong. °
Saclateni	206	Auriga	ζ	5.0	41	-0	165
Haedus	155	Auriga	η	5.1	41	0	165
Tania Australis	149	Ursa Major	μ	10.4	41	56	178
Chara	260	Canes Venatici	β	12.6	41	75	136
Almach	59	Andromeda	γ	2.1	42	-19	137
Alsciaukat / 31	261	Lynx	-	8.4	43	34	177
Tania Borealis	179	Ursa Major	λ	10.3	43	55	176
Almaaz	144	Auriga	ε	5.0	44	1.2	163
Alpherg	126	Pisces	η	1.5	45	-46	137
Misam	218	Perseus	κ	3.2	45	-11	147
Menkalinan	40	Auriga	β	6.0	45	10	167
La Superba	317	Canes Venatici	γ	12.8	45	72	127
Fawaris	131	Cygnus	δ	19.7	45	10	79
Deneb	19	Cygnus	α	20.7	45	2.0	84
Adhil	303	Andromeda	ξ	1.4	46	-17	129
Capella	6	Auriga	α	5.3	46	5	163
Intercrus / 41 Lynx[NA]	316	Ursa Major	-	9.5	46	46	174
Zavijava	197	Virgo	β	11.8	46	61	271
Xuange	259	Bootes	λ	14.3	46	65	87
Merga / 38	327	Bootes	-	14.8	46	60	80
85	>333	Hercules	ι	17.7	46	31	72
Alkaphrah	256	Ursa Major	κ	9.1	47	42	173
Ancha	257	Aquarius	θ	22.3	47	-49	54
Talitha	152	Ursa Major	ι	9.0	48	41	172
Taiyangshou	207	Ursa Major	χ	11.8	48	66	150
63	>333	Cygnus	-	21.1	48	0	89
Gal. 0, 90		Cygnus	-	21.2	48	0	90
Nembus / 51 / υ Persesu	196	Andromeda	-	1.6	49	-14	131
Alkaid / Benetnasch[NA]	37	Ursa Major	η	13.8	49	65	101
Mirfak	34	Perseus	α	3.4	50	-6	147
Eltanin	68	Draco	γ	17.9	51	29	79
Azelfafage	291	Cygnus	π¹	21.7	51	-1	95
Rastaban	113	Draco	β	17.5	52	33	80
23	133	Perseus	γ	3.1	53	-4	142

Table of Coordinates: by Decl-RA   **235** | P a g e

Common Name	#	Constellation	Gk	RA hrs	Decl. °	Glat. °	Glong. °
Fulu	204	Cassiopeia	ζ	0.6	54	-9	121
Phecda	78	Urda Major	γ	11.6	54	61	141
Alrakis	323	Draco	μ	17.1	54	37	82
Alruba	325	Draco	-	17.7	54	31	81
Mizar	65	Ursa Major	ζ	13.4	55	62	113
Alcor	241	Ursa Major	γ	13.6	55	61	113
Miram	214	Perseus	η	2.8	56	-3	138
Merak	74	Ursa Major	β	11.0	56	55	149
Alioth	32	Ursa Major	ε	12.9	56	61	122
Schedar	69	Cassiopeia	α	0.7	57	-6	121
Megrez	166	Ursa Major	δ	12.3	57	59	133
Grumium	211	Draco	ξ	17.9	57	30	85
Achird	180	Cassiopeia	η	0.8	58	-5	123
Caph	71	Cassiopeia	β	0.2	59	-3	117
Castula	287	Cassiopeia	υ²	0.9	59	-4	124
Edasich	164	Draco	ι	15.4	59	49	94
Ruchbah	99	Cassiopeia	δ	1.4	60	-2	127
Tsih[NA]	82	Cassiopeia	γ	0.9	61	-2	124
Muscida	169	Ursa Major	o	8.5	61	35	156
Athebyne / Alhibain[NA]	105	Draco	η	16.4	61	41	93
Dubhe	35	Ursa Major	α	11.1	62	51	143
Alderamin	81	Cepheus	α	21.3	63	9	101
Segin	170	Cassiopeia	ε	1.9	64	2	130
Thuban	203	Draco	α	14.1	64	51	111
Taiyi / 8	315	Draco	-	12.9	65	52	122
Aldhibah	154	Draco	ζ	17.1	66	35	96
Fafnir / 42	299	Draco	-	18.4	66	27	95
Tianyi / 7	318	Draco	-	12.8	67	50	124
Giausar	222	Draco	λ	11.5	69	+46	133
Alsafi	289	Draco	σ	19.5	70	22	101
Alfirk	159	Cepheus	β	21.5	71	14	108
Pherkad	141	Ursa Minor	γ	15.3	72	41	108
Dziban	284	Draco	ψ¹	17.7	72	31	103
Kochab	54	Ursa Minor	β	14.8	74	74	113
Tonatiuh	328	Camelopardalis	-	12.1	77	40	126
Errai	158	Cepheus	γ	23.7	78	15	119

Common Name	#	Constellation	Gk	RA hrs	Decl. °	Glat. °	Glong. °
Yildun	270	Ursa Minor	δ	17.5	87	28	119
Polaris	44	Ursa Minor	α	2.5	89	27	123

Table of Coordinates: by *Glat-Glong*

Common Name	#	Constellation	Gk	RA hrs	Decl. °	Glat. °	Glong. °
Giausar	222	Draco	λ	11.5	69	+46	133
HIP 4363	>333	Sculptor	-	0.9	-27	-89	114
Diphda	50	Cetus	β	0.7	-18	-81	111
Ankaa	77	Phoenix	α	0.4	-42	-74	320
Baten Kaitos	213	Ceteus	ζ	1.9	-10	-68	166
Fomalhaut	18	Piscis Austrinus	α	23.0	-30	-65	21
Angetenar	295	Eridanus	τ²	2.9	-21	-62	207
Wurren	245	Phoenix	ζ	1.1	-55	-62	298
Skat	163	Aquarius	δ	22.9	-16	-61	50
Acamar	124	Eridanus	θ¹	3.0	-40	-61	248
Dalim / 12 Eridanus	226	Fornax	α	3.2	-29	-59	225
Achernar	10	Eridanus	α	1.6	-57	-59	291
Mira	332	Ceteus	o	2.3	-3	-58	168
Tiaki	60	Grus	β	22.7	-47	-58	346
Alrescha	221	Pisces	α	2.0	3	-56	155
Revati	313	Pisces	ζ	1.2	8	-55	133
Azha	231	Eridanus	η	2.9	-8.9	-55	187
Alnair	30	Grus	α	22.1	-47	-53	350
Aldhanab	140	Grus	γ	21.9	-37	-51	6
Situla	309	Aquarius	κ	22.6	-4	-51	63
Torcular	265	Pisces	o	1.8	9	-51	145
Zibal	298	Eridanus	ζ	3.3	-9	-51	192
Fumalsamakah	279	Pisces	β	23.1	4	-50	79
Ancha	257	Aquarius	θ	22.3	47	-49	54
Kaffaljidhma	192	Ceteus	γ	2.7	14	-49	169
Ran	212	Eridanus	ε	3.5	-9	-48	196
-	123	Tucana	α	22.3	-61	-48	330
Algenib	118	Pegasus	γ	0.2	15	-47	109
Deneb Algedi	121	Capricornus	δ	21.8	-16	-46	38
Sadachbia	227	Aquarius	γ	22.4	-1	-46	62
Alpherg	126	Pisces	η	1.5	45	-46	137

Common Name	#	Constellation	Gk	RA hrs	Decl. °	Glat. °	Glong. °
Menkar	86	Cetus	α	3.0	4.1	-46	173
Nashira	205	Capricornus	γ	21.7	-17	-45	36
Zaurak	137	Eridanus	γ	4.0	-13	-45	205
Beemim	239	Eridanus	υ³	4.4	-34	-44	235
Sadalmelik	135	Aquarius	α	22.1	-1	-42	60
Homam	175	Pegasus	ζ	22.7	11	-41	79
Mesarthim	294	Aries	γ¹	1.9	19	-41	143
Theemin	220	Eridinus	υ	4.6	-31	-41	231
Bunda	297	Aquarius	ξ	21.6	-8	-40	46
Markab	85	Pegasus	α	23.1	15	-40	88
Sheratan	95	Aries	β	1.9	21	-40	142
-	117	Hydra	β	0.4	-77	-40	305
Biham	187	Pegasus	θ	22.2	6	-39	67
37	>333	Aries	o	2.7	15	-39	159
Sadalsuud	132	Aquarius	β	21.5	-6	-38	48
Beid	247	Eridanus	σ¹	4.2	-7	-38	199
Keid	276	Eridanus	o	4.3	-8	-38	201
Hamal	49	Aries	α	2.1	23	-36	145
Sceptrum / 53	244	Eridanus	-	4.6	-14	-36	211
Helvetios / 51	322	Pegasus	-	23.0	21	-35	90
Salm	285	Pegasus	τ	23.3	24	-35	98
Alkarab	274	Pegasus	υ	23.4	23	-35	99
Peacock	43	Pavo	α	20.4	-57	-35	341
Alpheratz	53	Andromeda	α	0.1	29	-33	112
Botein	269	Aries	δ	3.2	20	-32	163
Enif	76	Pegasus	ε	21.7	10	-31	66
Sadalbari	186	Pegasus	μ	22.8	25	-31	91
Mothallah	176	Triangulum	α	1.9	30	-31	139
Albali	216	Aquarius	ε	20.8	-9	-30	38
Kitalpha	235	Equuleus	α	21.3	5	-29	56
Scheat	80	Pegasus	β	23.1	28	-29	96
Bharani / 41	198	Aries	-	2.8	27	-29	153
Phact	98	Columba	α	5.7	-34	-29	239
Polaris Australis	320	Octans	σ	21.1	-89	-28	304
Mirach	55	Andromeda	β	1.2	36	-27	127
Lilii Borea / 39	282	Aries	-	2.8	29	-27	151
Nihal	115	Lepus	β	5.5	-21	-27	224

Table of Coordinates: by Glat-Glong

Common Name	#	Constellation	Gk	RA hrs	Decl. °	Glat. °	Glong. °
Wazn	153	Columba	β	5.8	-36	-27	241
Dabih	147	Capricornus	β¹	20.4	-15	-26	29
Terebellum	292	Sagittarius	ω	19.9	-26	-25	15
Algedi	194	Capricornus	α²	20.3	-13	-25	31
Alshat	296	Capricornus	ν	20.3	-13	-25	31
Matar	133	Pegasus	η	22.7	30	-25	93
4	142	Triangulum	β	2.2	35	-25	141
Cursa / Kursa[NA]	111	Eridanus	β	5.1	-5	-25	205
Rigel	7	Orion	α	5.2	-8	-25	209
Arneb	89	Lepus	α	5.5	-18	-25	221
Canopus	2	Carina	α	6.4	-53	-25	261
Celaeno / 16	319	Taurus	-	3.7	24	-24	166
Electra / 17	209	Taurus	-	3.7	24	-24	166
Maia / 20	229	Taurus	-	3.8	24	-24	166
Taygeta / 19	266	Taurus	q	3.8	24	-24	166
Merope / 23	255	Taurus	-	3.8	24	-24	167
Prima Hyadum	201	Taurus	γ	4.3	16	-24	179
Elkurud	306	Columba	θ	6.1	-37	-24	244
Arkab Posterior	264	Sagittarius	β²	19.4	-45	-24	353
Arkab Prior	237	Sagittarius	β¹	19.4	-44	-24	354
Asterope / 21	326	Taurus	-	3.8	25	-23	166
Alcyone	120	Taurus	η	3.8	24	-23	167
Atlas / 27	199	Taurus	-	3.8	24	-23	167
Pleione / 28	310	Taurus	-	3.8	24	-23	167
Tabit	157	Orion	π³	4.8	7	-23	191
Rukbat	238	Sagittarius	α	19.4	-41	-23	358
Musica / 18	321	Delphinus	-	21.0	11	-22	59
Secunda Hyadum	217	Taurus	δ	4.4	18	-22	178
Chamukuy	210	Taurus	θ	4.5	16	-22	180
Veritate / 14	314	Andromeda	-	23.5	39	-21	106
Titawin	251	Andromeda	υ	1.6	41	-21	132
Altaleban[NA]	157	Puppis	τ	6.8	-51	-21	260
Ain	190	Taurus	ε	4.5	19	-20	178
Aldebaran	14	Taurus	α	4.6	17	-20	181
Hatysa	114	Orion	ι	5.6	-6	-20	210
Meridiana	252	Corona	α	19.2	-38	-20	359

Common Name	#	Constellation	Gk	RA hrs	Decl. °	Glat. °	Glong. °
		Australis					
Almach	59	Andromeda	γ	2.1	42	-19	137
Saiph	56	Orion	κ	5.8	-10	-19	215
Furud	143	Canis Major	ζ	6.3	-30	-19	238
65	161	Aquila	θ	20.2	-1	-18	42
Atik	224	Perseus	o	3.7	32	-18	160
Mintaka	70	Orion	δ	5.5	-0	-18	204
Aldulfin	246	Delphinus	ε	20.6	11	-17	55
Adhil	303	Andromeda	ξ	1.4	46	-17	129
44	119	Perseus	ζ	3.9	32	-17	162
Alnilam	29	Orion	ε	5.6	-1	-17	205
Alnitak	31	Orion	ζ	5.7	-2	-17	207
Rotanev	200	Delphinus	β	20.6	15	-16	59
Bellatrix	26	Orion	γ	5.4	6	-16	197
Algol	58	Perseus	β	3.1	41	-15	3
Ascella	91	Sagittarius	ζ	19.0	-30	-15	7
Sualocin	215	Delphinus	α	20.7	16	-15	60
94	>333	Taurus	τ	4.7	23	-15	177
Atria	41	Triangulum Australe	α	16.8	-69	-15	321
Nembus / 51 / υ Persesu	196	Andromeda	-	1.6	49	-14	131
Mirzam	46	Canis Major	β	6.4	-18	-14	226
Miaplacidus	28	Carina	β	9.2	-70	-14	286
Nunki	51	Sagittarius	σ	18.9	-26	-13	10
Albaldah	125	Sagittarius	π	19.2	-21	-13	16
55	>333	Aquila	η	19.9	1	-13	41
Menkib	240	Perseus	ξ	4.0	36	-13	160
Avior	38	Carina	ε	8.4	-59	-13	274
Meissa / Heka[NA]	173	Orion	λ	5.6	10	-12	195
Ainalrami	302	Sagittarius	ν	18.9	-23	-11	13
Alshain	208	Aquila	β	19.9	6	-11	46
Misam	218	Perseus	κ	3.2	45	-11	147
Adhara	23	Canis Major	ε	7.0	-29	-11	240
-	102	Puppis	π	7.3	-37	-11	249
-	119	Ara	β	17.4	-55	-11	335
Cervantes	312	Ara	μ	17.7	-52	-11	341

Common Name	#	Constellation	Gk	RA hrs	Decl. °	Glat. °	Glong. °
Libertas	293	Aquila	ξ	19.9	8	-10	48
45	133	Perseus	ε	4.0	40	-10	157
Unurgunite	184	Canis Major	σ	7.0	-28	-10	239
Kaus Australis	33	Sagittarius	ε	18.4	-34	-10	359
Altair	12	Aquila	α	19.8	9	-9	48
Fulu	204	Cassiopeia	ζ	0.6	54	-9	121
Betelgeuse	9	Orion	β	5.9	7	-9	200
Sirius	1	Canis Major	α	6.8	-17	-9	227
-	133	Carina	υ	9.8	-65	-9	285
-	123	Triangulum Australe	γ	15.3	-69	-9	316
-	119	Ara	α	17.5	-50	-9	341
Wezen	36	Canis Major	δ	7.1	-26	-8	238
Regor^NA	33	Vela	γ	8.2	-47	-8	263
Kaus Media	103	Sagittarius	δ	18.3	-29.8	-7.2	3.0
Tarazed	104	Aquila	γ	19.8	11	-7	49
Aludra	82	Canis Major	η	7.6	-29	-7	243
Alsephina	48	Vela	δ	8.7	-55	-7	272
Aspidiske / Turais^NA	62	Carina	ι	9.3	-59	-7	279
-	121	Triangulum Australe	β	15.9	-63	-7	322
Kaus Borealis	116	Sagittarius	λ	16.5	-25.4	-6.5	7.7
Aljanah / Gienah Cygni^NA	84	Cygnus	ε	20.8	34	-6	76
Schedar	69	Cassiopeia	α	0.7	57	-6	121
Mirfak	34	Perseus	α	3.4	50	-6	147
Hassaleh	101	Auriga	ι	5.0	33	-6	171
Tianguan	136	Taurus	ζ	5.6	21	-6	186
-	101	Musca	α	12.6	-69	-6	301
Sargas	39	Scorpius	θ	17.7	-43	-6	347
Alnasl	138	Sagittarius	γ²	18.1	-30	-5	1
Achird	180	Cassiopeia	η	0.8	58	-5	123
Naos	63	Puppis	ζ	8.1	-40	-5	256
-	107	Carina	θ	10.7	-64	-5	290
Girtab^NA	81	Scorpius	κ	17.6	-39	-5	351
Castula	287	Cassiopeia	υ²	0.9	59	-4	124

Common Name	#	Constellation	Gk	RA hrs	Decl. °	Glat. °	Glong. °
23	133	Perseus	γ	3.1	53	-4	142
Elnath	27	Taurus	β	5.4	29	-4	178
Muliphein	253	Canis Major	γ	7.1	-16	-4	228
Caph	71	Cassiopeia	β	0.2	59	-3	117
Miram	214	Perseus	η	2.8	56	-3	138
Markeb	83	Vela	κ	9.4	-55	-3	276
Polis	225	Sagittarius	μ	18.2	-21	-2	10
Sham	271	Sagitta	α	19.6	18	-2	55
Tsih^{NA}	82	Cassiopeia	γ	0.9	61	-2	124
Ruchbah	99	Cassiopeia	δ	1.4	60	-2	127
Lesath	102	Scorpius	υ	17.5	-37	-2	351
Shaula	25	Scorpius	λ	17.6	-37	-2	352
Azelfafage	291	Cygnus	π¹	21.7	51	-1	95
Rigel Kentaurus	3	Centaurus	α	14.7	-60	-1	316
Gal. 0, 0		Sagittarius	-	17.8	-29	0	0
63	>333	Cygnus	-	21.1	48	0	89
Gal. 0, 90		Cygnus	-	21.2	48	0	90
Fuyue	156	Scorpius	η	5.1	14	0	165
Haedus	155	Auriga	η	5.1	41	0	165
Saclateni	206	Auriga	ζ	5.0	41	-0	165
Gal. 0, 180		Taurus	-	5.8	29	0	180
Gal. 0, 270		Vela	-	9.2	-48	0	270
Acrux	13	Crux	α	12.4	-63	-0	300
Alya	286	Serpens	θ¹	18.9	4.2	1	37
Azmidi	178	Puppis	ξ	7.8	-25	1	241
Hadar	11	Centaurus	β	14.1	-60	1	312
Almaaz	144	Auriga	ε	5.0	44	1.2	163
Sadr	66	Cygnus	γ	20.4	40	2	78
Deneb	19	Cygnus	α	20.7	45	2.0	84
Segin	170	Cassiopeia	ε	1.9	64	2	130
Ginan	195	Crux	ε	12.4	-60	2	299
Okab	139	Aquila	ζ	19.1	14	3	47
Anser	277	Vulcan	α	19.5	25	3	59
Propus	165	Gemini	η	16.2	23	3	189
Suhail	67	Vela	λ	9.1	-43	3	266
Tejat	123	Gemini	μ	6.4	31	4	190
Imai	108	Crux	δ	12.3	-59	4	298

Common Name	#	Constellation	Gk	RA hrs	Decl. °	Glat. °	Glong. °
Pipirima	193	Scorpius	μ²	16.9	-38	4	346
Xamidimura	142	Scorpius	μ¹	16.9	-38	4	346
Deneb El Okab^NA	245	Aquila	ε	19.0	15	5	47
Albireo	146	Cygnus	β¹	19.5	28	5	62
Muphrid	100	Bootes	η	13.9	18	5	73
Capella	6	Auriga	α	5.3	46	5	163
Alhena / Almeisan^NA	42	Gemini	γ	6.6	16	5	197
Alzirr	168	Gemini	ξ	6.8	13	5	201
Tureis	119	Puppis	ρ	8.1	-24	5	243
Gacrux	24	Crux	γ	12.5	-57	6	300
Kurhah	263	Cepheus	ξ	22.1	38	7	106
Mahasim	97	Auriga	θ	6.0	37	7	174
Denebola	61	Leo	β	11.8	15	7`	251
Larawag	73	Scorpius	ε	16.8	-34	7	349
Guniibuu / 36	311	Ophiuchus	-	17.3	-27	7	358
Alderamin	81	Cepheus	α	21.3	63	9	101
-	102	Vela	μ	11	-49	9	283
-	74	Centaurus	ε	13.7	-53	9	310
Fawaris	131	Cygnus	δ	19.7	45	10	79
Menkalinan	40	Auriga	β	6.0	45	10	167
Mebsuta	148	Gemini	ε	6.7	25	10	189
Kakkab^NA	75	Lupus	α	14.7	-43	11	322
Mekbuda	242	Gemini	ζ	7.1	21	12	196
Gomeisa	129	Canis Minor	β	7.5	8	12	209
-	91	Centaurus	δ	12.1	-51	12	296
-	114	Lupus	γ	15.6	-41	12	333
Sulafat	162	Lyra	γ	19.0	33	13	63
Aladfar	275	Lyra	η	19.2	39	13	71
Procyon	8	Canis Minor	α	7.7	5	13	214
Paikauhale	117	Scorpius	τ	16.6	-28	13	352
Sabik	79	Ophiuchus	η	17.2	-16	14	7
Alfirk	159	Cepheus	β	21.5	71	14	108
Muhlifain^NA	62	Centaurus	γ	12.7	-49	14	301
Kekouan^NA	111	Lupus	β	15.0	-43	14	326
Antares	16	Scorpius	α	16.5	-26	15	-8

Table of Coordinates: by Glat-Glong

Common Name	#	Constellation	Gk	RA hrs	Decl. °	Glat. °	Glong. °
Sheliak	188	Lyra	β	18.8	33	15	63
Errai	158	Cepheus	γ	23.7	78	15	119
Wasat	185	Gemini	δ	7.3	22	16	196
Cebalrai	110	Ophiuchus	β	17.6	5	17	29
-	78	Centaurus	η	14.6	-42	17	323
Alniyat	130	Scorpius	σ	16.4	-26	17	351
Iklil	228	Scorpius	ρ	15.9	-29	18	345
Vega	5	Lyra	α	18.6	38	19	67
Fang	128	Scorpius	π	16.0	-26	20	347
Alsafi	289	Draco	σ	19.5	70	22	101
Rasalhague	57	Ophiuchus	α	17.6	13	23	36
Altais	150	Draco	δ	19.2	40	23	99
Castor	45	Gemini	α	7.6	32	23	187
Pollux	17	Gemini	β	7.8	28	23	192
Tarf / Altarf[NA]	191	Cancer	β	8.3	9	23	214
Dschubba	72	Scorpius	δ	16.0	-23	23	350
Jabbah	280	Scorpius	ν	16.2	-19	23	355
13	86	Ophiuchus	ζ	16.6	-11	24	6
Jishui	304	Gemini	o	7.7	35	24	185
Menkent	52	Centaurus	θ	14.1	-36	24	319
Acrab / Graffias[NA]	87	Scorpius	β¹	16.1	-20	24	353
Minchir	278	Hydra	σ	8.6	3	25	223
Tegmine	290	Cancer	ζ¹	8.2	18	26	205
-	110	Centaurus	ι	13.3	-37	26	309
Fafnir / 42	299	Draco	-	18.4	66	27	95
Polaris	44	Ursa Minor	α	2.5	89	27	123
Felis	305	Hydra	-	9.9	-19	27	255
Rasalgethi	171	Hercules	α¹	17.2	14	28	36
Maasym	273	Hercules	λ	17.5	26	28	49
Yildun	270	Ursa Minor	δ	17.5	87	28	119
Eltanin	68	Draco	γ	17.9	51	29	79
Ashlesha	183	Hydra	ε	8.8	6	29	221
Alphard	47	Hydra	α	9.5	-9	29	241
Zhang	254	Hydra	υ¹	9.9	-15	29	251
Brachium	161	Libra	σ	15.1	-25	29	337
Grumium	211	Draco	ξ	17.9	57	30	85

Table of Coordinates: by Glat-Glong   **245** | P a g e

Common Name	#	Constellation	Gk	RA hrs	Decl. °	Glat. °	Glong. °
Piautos	329	Cancer	λ	8.3	24	30	199
Yed Posterior	160	Ophiuchus	ε	16.3	-4.7	31	9
Sarin	151	Hercules	δ	17.3	25	31	47
85	>333	Hercules	ι	17.7	46	31	72
Alruba	325	Draco	-	17.7	54	31	81
Dziban	284	Draco	ψ¹	17.7	72	31	103
Yed Prior	106	Ophiuchus	δ	16.2	-3.7	32	9
Marfik	223	Opheuchus	λ	16.5	2	32	17
Meleph	331	Cancer	ε	8.7	20	32	206
24	>333	Libra	ι	15.2	-20	32	343
Zubenelhakradi	234	Libra	γ	15.6	-15	32	352
Rastaban	113	Draco	β	17.5	52	33	80
Asellus Asutralis	236	Cancer	δ	8.7	18	33	208
Mimosa	20	Crux	β	12.8	-60	33	303
67	154	Hercules	π	17.3	37	34	61
Alsciaukat / 31	261	Lynx	-	8.4	43	34	177
Asellus Borealis	288	Cancer	γ	8.7	21	34	204
Acubens	262	Cancer	α	9.0	12	34	217
Aldhibah	154	Draco	ζ	17.1	66	35	96
Muscida	169	Ursa Major	o	8.5	61	35	156
Ukdah	233	Hydra	ι	9.7	-1	36	236
Alrakis	323	Draco	μ	17.1	54	37	82
Alkes	249	Crater	α	11	-18	37	269
Alchiba	243	Corvus	α	12.1	-25	37	291
Copernicus / 55	330	Cancer	-	8.9	28	38	197
Zubenelgenubi	109	Libra	α²	14.8	-16	38	340
Cujam	283	Hercules	ω	16.4	14	39	29
Kraz	96	Corvus	β	12.6	-23	39	298
Zubene-schamali	93	Libra	β	15.3	-9	39	352
Kornephoros	112	Hercules	β	16.5	21	40	39
RutilicusNA	133	Hercules	ζ	16.7	32	40	53
Tonatiuh	328	Camelopardalis	-	12.1	77	40	126
Nahn	324	Cancer	ξ	9.2	22	40	206
Athebyne / AlhibainNA	105	Draco	η	16.4	61	41	93
Pherkad	141	Ursa Minor	γ	15.3	72	41	108

Table of Coordinates: by Glat-Glong   **246** | P a g e

Common Name	#	Constellation	Gk	RA hrs	Decl. °	Glat. °	Glong. °
Talitha	152	Ursa Major	ι	9.0	48	41	172
Alkaphrah	256	Ursa Major	κ	9.1	47	42	173
Subra	189	Leo	o	9.7	10	42	225
Unukalhai	94	Serpens	α	15.7	6	44	14
Marsic	307	Hercules	κ	16.1	17	44	31
Khambalia	281	Virgo	λ	14.3	-13	44	333
38	>333	Lynx	-	9.3	37	45	187
40	153	Lynx	α	9.4	34	45	190
Alterf	269	Leo	λ	9.5	23	45	207
Gienah / Gienah Ghurab[NA]	90	Corvus	γ	12.3	-17	45	291
Intercrus / 41 Lynx[NA]	316	Ursa Major	-	9.5	46	46	174
Algorab	134	Corvus	δ	12.5	-17	46	295
107	>333	Virgo	μ	14.7	-6	47	347
Gudja	250	Serpens	κ	15.6	18	48	30
17	137	Leo	ε	9.8	24	48	207
Kang	258	Virgo	κ	14.2	-10	48	334
Edasich	164	Draco	ι	15.4	59	49	94
Regulus	21	Leo	α	10.1	12	49	226
Tianyi / 7	318	Draco	-	12.8	67	50	124
Rasalas	230	Leo	μ	9.9	26	50	204
Thuban	203	Draco	α	14.1	64	51	111
Dubhe	35	Ursa Major	α	11.1	62	51	143
Spica	15	Virgo	α	13.4	-11	51	316
Syrma	248	Virgo	ι	14.3	-6	51	338
Taiyi / 8	315	Draco	-	12.9	65	52	122
Elgafar	300	Virgo	φ	14.5	-2	53	345
109	>333	Virgo	-	14.8	2	53	355
Alphecca / Gemma[NA]	64	Corona Borealis	α	15.6	27	54	42
-	122	Hydrus	α	2.0	-62	54	289
Merak	74	Ursa Major	β	11.0	56	55	149
Tania Borealis	179	Ursa Major	λ	10.3	43	55	176
Adhafera	177	Leo	ζ	10.3	23	55	210
Algieba	92	Leo	γ¹	10.3	20	55	217
Nusakan	202	Corona	β	15.5	29	56	46

Table of Coordinates: by Glat-Glong   **247** | P a g e

Common Name	#	Constellation	Gk	RA hrs	Decl. °	Glat. °	Glong. °
		Borealist					
Alkalurops	267	Bootes	μ^2	15.4	37	56	60
Tania Australis	149	Ursa Major	μ	10.4	41	56	178
49	180	Bootes	δ	15.3	33	58	53
Megrez	166	Ursa Major	δ	12.3	57	59	133
Nekkar	183	Bootes	β	15.0	40	60	68
Merga / 38	327	Bootes	-	14.8	46	60	80
Heze	172	Virgo	ζ	13.6	-1	60	325
30	>333	Bootes	ζ	14.7	14	61	11
Alcor	241	Ursa Major	γ	13.6	55	61	113
Alioth	32	Ursa Major	ϵ	12.9	56	61	122
Phecda	78	Urda Major	γ	11.6	54	61	141
Zavijava	197	Virgo	β	11.8	46	61	271
Zaniah	232	Virgo	η	12.3	-1	61	286
Porrima	107	Virgo	γ	12.7	-1	61	298
Mizar	65	Ursa Major	ζ	13.4	55	62	113
Chalawan / 47	308	Ursa Major	-	11.0	40	63	176
Praecipua / 46	219	Leo Minor	-	10.9	34	64	190
Izar	75	Bootes	ϵ	14.7	27	65	39
Xuange	259	Bootes	λ	14.3	46	65	87
Alkaid / Benetnasch[NA]	37	Ursa Major	η	13.8	49	65	101
Chertan	167	Leo	θ	11.2	15.4	65	235
Seginus	145	Bootes	γ	14.5	38	66	67
Taiyangshou	207	Ursa Major	χ	11.8	48	66	150
Minelauva	174	Virgo	δ	12.9	3	66	306
Zosma	88	Leo	δ	11.2	21	67	224
Arcturus	4	Bootes	α	14.3	19	69	15
Alula Borealis	182	Ursa Major	ν	11.3	33	69	191
Alula Australis	272	Ursa Major	ξ	11.3	32	69	195
La Superba	317	Canes Venatici	γ	12.8	45	72	127
Kochab	54	Ursa Minor	β	14.8	74	74	113
Vindemaitrix	122	Virgo	ϵ	13.0	11	74	312
Chara	260	Canes Venatici	β	12.6	41	75	136
Cor Caroli	127	Canes Venatici	α^2	12.9	38	79	118
Diadem	301	Coma Berenices	α	3.2	18	79	328

Common Name	#	Constellation	Gk	RA hrs	Decl. °	Glat. °	Glong. °
HIP 62752 A	>333	Coma Berenices	-	12.9	27	90	120

Table of Near Galaxies, Meteor Showers & Nebulas

Object	RA	Decl	
Galaxies (visible by eye)			**Constellation**
Andromeda	0.7	41	Andromeda
Large Magellanic Cloud	5.4	-70°	Dorado / Mensa
Small Magellanic Cloud	0.9	-73	Hydrus / Tucana
Meteor Shower (popular) **by Peaks**			**Peak**
Quadrantids in Bootes	15.3	50	Jan. 4±
Lyrids	18.1	34	Apr. 22±
Eta Aquariids	22.5	-1	May 6±
Alpha Capricornids	20.3	-10	Jul. 29±
Delta Aquariids	22.7	-16	Jul. 29±
Perseids	3.2	58	Aug. 13±
Draconids	17.5	54	Oct. 7±
Southern Taurids	2.1	9	Oct 10±
Orionids	6.3	15	Oct. 22±
Northern Taurids	3.9	23	Nov. 12±
Leonids	10.1	22	Nov. 17±
Geminids	7.5	32	Dec. 14±
Ursids	14.3	75	Dec 22±
Nebula (nearly visible)			**Constellation**
Coalsack (dark)	12.8	-63	Crux
Crab Nebula (dead supernova)	5.6	22	Taurus
Dumbbell Nebula (planetary)	20.0	23	Vulpecula
Eta Carinae Nebula (star formation)	10.8	-60	Carina
Orion Nebula (star formation)	5.6	-5	Orion
Ring Nebula	18.9	33	Lyra

Object	RA	Decl	
(planetary)			

Table of All Constellations, their Stories

88 Constellations (1) CE-RA0h-centred (from Stellarium)

The above sky view is an identical field-of-view as the rear half-shell of the **CS** in figure **22 Celestial Sphere showing the Milky Way band** as viewed from *inside* the **CS** by an observer on Earth.

The following figure is a simplified version of the above rendering, containing just constellation labels to make it easy to find and then looking for it in the above where its detailed linework can be examined. But, it also has some goodies: constellations of antiquity are in white, while the modern ones are in grey; Zodiac constellations are underlined; the **CE** and the Ecliptic are shown.

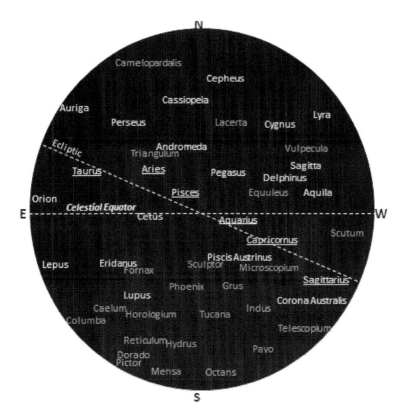

89 Constellations (1) CE-RA0h-centred

Since screenshots are never as sharp as a drawing like above, recourse can be had to a constellation-by-constellation look at its linework: At the IAU official site at https://www.iau.org/public/themes/constellations/, under the "Charts and tables" section. Clicking on any constellation's chart icon brings up a crisp image of its linework, boundary, star names, as well as any galaxies and nebulae that are within its field-of-view, plus neighbouring constellations (but without their object labels). Here is an example, that of Orion the Hunter:

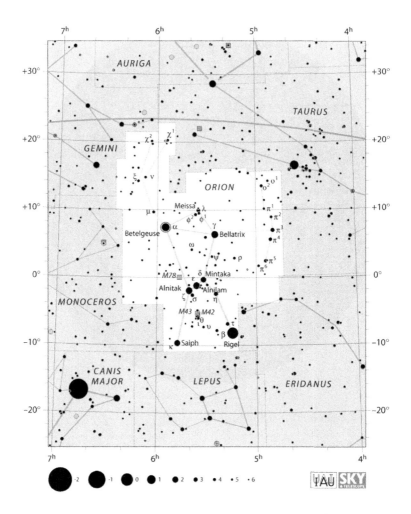

90 Chart of Orion the Hunter (by the IAU)

91 Constellations (2) CE-RA12h-centred (from Stellarium)

The above sky view is an identical field-of-view as the front half-shell of the **CS** in figure **22 Celestial Sphere showing the Milky Way band** as viewed from *inside* the **CS** by an observer on Earth. Like the previous pair of figures – a graphic one with constellation linework, and a simplified one with just labels – below is a similarly simplified version of the above graphic one.

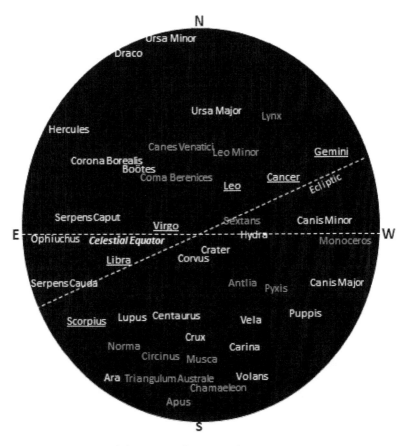

92 Constellations (2) CE-RA12h-centred

Below is a complete list of the 88 IAU (International Astronomical Union) constellations, with the vintage ones of millennia past bolded.

Constellation ranks are superscripted, this ranking being based on the coverage sizes of their arbitrary boundaries.

Constellation	Fig., Loc.	Neighbours, <u>The Story</u>
Andromeda[19]	Fig. **89** NEsector *RA1h±,*	Between Pegasus & Perseus. <u>The Chained Woman</u> Daughter of Cepheus & Cassiopeia,

Constellation	Fig., Loc.	Neighbours, **The Story**
	Decl35°±	wife of Perseus. Chained by her father to appease sea monster Cetus, rescued by Perseus.
Antlia[62]	Fig. **92** SWsector *RA10h±,* *Decl-35°±*	Air Pump.
Apus[67]	Fig. **92** SEsector *RA16h±,* *Decl-75°±*	Bird of Paradise.
Aquarius[10]	Fig. **89** SWsector *RA22h±,* *Decl-10°±*	Between fellow Zodiacs, Capricornus & Pisces, its top just cutting the *CE* close to *RA22h*. <u>The Water Carrier</u> Babylonian god Ea, associated with destructive floods but in Egypt with the necessary spring flooding of the Nile.
Aquila[22]	Fig. **89** NWsector *RA20h±,* *Decl5°±*	Between Capricornus & Hercules, its lower parts cutting the *CE* near *RA20h*. Its brightest star Altair is one corner of the well-known asterism Summer Triangle. <u>The Eagle</u> Delivers thunder from Greek god Zeus (or Roman god Jupiter).
Ara[63]	Fig. **92** SEsector *RA17h±,* *Decl-55°±*	Between Pavo & Scorpius. <u>The Altar</u> Where the early gods made their sacrifices.
Aries[39]	Fig. **89** NEsector *RA2h±,* *Decl25°±*	Between fellow Zodiacs, Pisces & Taurus. <u>The Ram</u> Egyptian god Amun-Ra (or Amon-

Constellation	Fig., Loc.	Neighbours, <u>The Story</u>
		Ra, or Amen-Ra) with a ram's head.
Auriga[21]	Fig. **89** NEsector *RA5h±,* *Decl40°±*	Between Lynx & Perseus. Its 1st Order *Magnitude* star Capella is one corner of the Winter Hexagon asterism. <u>The Charioteer or Goatherd</u> Greek hero Erichthonius, king of Athens.
Bootes[13]	Fig. **92** NEsector *RA15h±,* *Decl30°±*	Between Hercules & Virgo. Resembles a kite. Its brightest star is one corner of the Spring Triangle asterism, which in turn is the lower half of the Great Diamond. asterism (see constellation Canes Venatici). The Quadrantids meteor shower occurs near its northern boundary. <u>The Herdsman</u> Babylonian god Enlil, patron of agriculture.
Caelum[81]	Fig. **89** SEsector *RA5h±,* *Decl-40°±*	Chisel.
Camelopardalis[18]	Fig. **89** NEsector *RA6h±,* *Decl70°±*	Giraffe.
Cancer[31]	Fig. **92** NWsector *RA9h±,* *Decl20°±*	Between fellow Zodiacs, Gemini & Leo. <u>The Crab</u> The crab that bit Hercules. Eponym of Tropic of Cancer (Lat. 23.5°) when in ancient days the

Constellation	Fig., Loc.	Neighbours, <u>The Story</u>
		apparent Sun was at the highest *Declination* in this constellation, but nowadays it is in Gemini.
Canes Venatici[38]	Fig. **92** NEsector *RA13h±, Decl40°±*	Hunting Dogs. Its star Cor Caroli is the top of the Great Diamond asterism.
Canis Major[43]	Fig. **92** SWsector *RA7h±, Decl-20°±*	Between Orion & Puppis. Its 1st Order *Magnitude* star Sirius is one corner of the Winter Hexagon asterism as well as of the Winter Triangle asterism. <u>The Greater Dog</u> A gift from Greek god Zeus to Europa (eponym for Europe) whom he had abducted. Also known as one of Orion's hunting dogs.
Canis Minor[71]	Fig. 92 NWsector *RA8h±, Decl5°±*	Between Hydra & Orion. Its 1st Order *Magnitude* star Procyon is one corner of the Winter Hexagon asterism as well as of the Winter Triangle asterism. <u>The Lesser Dog</u> Turned into stone by Greek god Zeus. Also known as one of Orion's hunting dogs.
Capricornus[40]	Fig. **89** SWsector *RA21h±, Decl-20°±*	Between fellow Zodiacs, Aquarius & Sagittarius. <u>The Sea-Goat</u> Part fish, part goat. Its broken horn was turned into Cornucopia, the horn of plenty.
Carina[34]	Fig. **92** SWsector	South of its related constellations Puppis & Vela. Its stars Aspidiske

Constellation	Fig., Loc.	Neighbours, The Story
	RA9h±, *Decl-60°±*	and Avior are part of the False Cross asterism. The Keel Of the old constellation Argo Navis (the ship of Jason and the Argonauts, in search of the Golden Fleece) which was divided into three constellations.
Cassiopeia[25]	Fig. **89** NEsector *RA1h±,* *Decl60°±*	Between Cepheus & Perseus, in the shape of a queen's crown, easy to recognize! Queen of Ethiopia She bragged about her daughter Cassiopeia's beauty. In retaliation, god Poseidon ordered the daughter to be made a prey to sea monster Cetus.
Centaurus[9]	Fig. **92** SEsector *RA13h±,* *Decl-50°±*	Between Crux & Lupus. It is one of only three constellations that has two 1st Order **Magnitude** stars. The Centaur Half human, half horse, named Chiron, mentor to heroes like Hercules and Jason. Note: Sagittarius also represents a centaur albeit a different one.
Cepheus[27]	Fig. **89** NWsector *RA22h±,* *Decl70°±*	Between **CNP** & Cygnus. King of Ethiopia Husband of Queen Cassiopeia, father of Andromeda.
Cetus[4]	Fig. **89** SEsector *RA2h±,* *Decl-10°±*	Between Eridanus & Pisces, its upper portions cuts the **CE** at *RA2.5h±.* The Sea Monster (or Whale) Slayed by Perseus, in the rescue of

Constellation	Fig., Loc.	Neighbours, **The Story**
		Andromeda. It is in the water-theme area of the sky viz. Eridanus the River, Pisces the Fishes, Aquarius the Water Carrier.
Chamaeleon[79]	Fig. **92** SWsector *RA11h±, Decl-80°±*	Chameleon.
Circinus[85]	Fig. **92** SEsector *RA15h±, Decl-65°±*	Draftsman's Compass cf. constellation Pyxis (Mariner's Compass).
Columba[54]	Fig. **89** SEsector *RA6h±, Decl-35°±*	Dove.
Coma Berenices[42]	Fig. **92** NEsector *RA13h±, Decl25°±*	Hair of Queen Berenice II of Egypt. Home of the Galactic North Pole!
Corona Australis[80]	Fig. **89** SWsector *RA19h±, Decl-40°±*	Three quarters surrounded by Sagittarius. <u>The Southern Crown or Wreath</u> Fallen off the head of the centaur (Sagittarius).
Corona Borealis[73]	Fig. **92** NEsector *RA16h±, Decl30°±*	Between Bootes & Hercules. <u>The Northern Crown</u> Given in marriage by god Dionysus to princess Ariadne, daughter of Minos (king of Crete).
Corvus[70]	Fig. **92** SEsector *RA12h±, Decl-20°±*	Between Hydra & Virgo. <u>The Crow/Raven</u> Sits on the water serpent Hydra. It told Apollo of his lover's infidelity, Apollo blackening its feathers in

Constellation	Fig., Loc.	Neighbours, <u>The Story</u>
		rage.
Crater[53]	Fig. **92** SWsector *RA11h±,* *Decl-20°±*	Between Corvus & Hydra. <u>The Crater (*krater*, cup in Greek)</u> Apollo's cup on Hydra' back.
Crux[88]	Fig. **92** SEsector *RA12h±,* *Decl-60°±*	Between Centaurus & Vela, it is the smallest constellation. It is one of only three constellations that has two 1st Order **Magnitude** stars. <u>The Southern Cross</u> Its distinct asterism is self-explanatory.
Cygnus[16]	Fig. **89** NWsector *RA20h±,* *Decl40°±*	Between Pegasus & Lyra. Contains the asterism Northern Cross. Its brightest star Deneb is one corner of the well-known asterism Summer Triangle. The Galactic 90° lies in it. <u>The Swan</u> This is king Cygnus made into a swan.
Delphinus[69]	Fig. **89** NWsector *RA21h±,* *Decl15°±*	Between Aquila & Pegasus. <u>The Dolphin</u> It persuaded Amphitrite to return to god Poseidon.
Dorado[72]	Fig. **89** SEsector *RA5h±,* *Decl-60°±*	Dolphin.
Draco[8]	Fig. **92** NEsector *RA17h±,* *Decl65°±*	Between Hercules & Ursa Minor. <u>The Dragon</u> The guardian of the Golden Apples, being the 11th of the 12 Labours of Hercules who slays it.
Equuleus[87]	Fig. **89**	Little Horse.

Constellation	Fig., Loc.	Neighbours, **The Story**
	NWsector *RA21h±, Decl10°±*	
Eridanus[6]	Fig. **89** SEsector *RA4h±, Decl-25°±*	Between Cetus & Orion. The River Greek name of River Po in Italy.
Fornax[41]	Fig. **89** SEsector *RA3h±, Decl-30°±*	Furnace.
Gemini[30]	Fig. **92** NWsector *RA7h±, Decl25°±*	Between fellow Zodiacs, Cancer & Taurus. Its 1st Order **Magnitude** star Pollux is one corner of the Winter Hexagon asterism. The Twins (half-brothers) One mother, Spartan queen Leda, Pollux' father, god Zeus, Castor's father, Spartan king Tyndareus, hence the latter a mortal, upon whose death the former begs Zeus to immortalize him, and uniting them in the heavens.
Grus[45]	Fig. **89** SWsector *RA22h±, Decl-45°±*	Crane (bird).
Hercules[5]	Fig. **92** NEsector *RA17h±, Decl30°±*	Between Bootes & Lyra, contains the Keystone asterism. The Strong He of the 12 Labours, son of god Zeus, seen in the constellation as kneeling in prayer.
Horologium[58]	Fig. **89** SEsector	Clock.

Constellation	Fig., Loc.	Neighbours, The Story
	RA3h±, *Decl-55°±*	
Hydra[1]	Fig. **92** SWsector *RA11h±,* *Decl-20°±*	Between Centaurus & Libra, it is the largest constellation. The Water Snake (female) Slayed by Hercules as the 2nd of his 12 Labours.
Hydrus[61]	Fig. **89** SEsector *RA2h±,* *Decl-70°±*	Water snake (male).
Indus[49]	Fig. **89** SWsector *RA21h±,* *Decl-55°±*	Indus (river).
Lacerta[68]	Fig. **89** NWsector *RA22h±,* *Decl45°±*	Lizard.
Leo Minor[64]	Fig. **92** NWsector *RA10h±,* *Decl35°±*	Smaller Lion.
Leo[12]	Fig. **92** NWsector *RA11h±,* *Decl15°±*	Between fellow Zodiacs, Cancer & Virgo. Its star Denebola is one corner of the asterism Spring Triangle, which in turn is the lower half of the Great Diamond asterism (see constellation Canes Venatici). The Lion Killed by Hercules as the 1st of his 12 Labours.
Lepus[51]	Fig. **89** SEsector	Between Columba & Orion. The Hare

Constellation	Fig., Loc.	Neighbours, The Story
	RA6h±, *Decl-15°±*	Hunted by Orion's hunting dogs, Canis Major & Minor.
Libra[29]	Fig. **92** SEsector *RA15h±,* *Decl-15°±*	Between fellow Zodiacs, Scorpius and Virgo. The Weighing Scales Representing Justice.
Lupus[46]	Fig. **92** SEsector *RA15h±,* *Decl-40°±*	Between Centaurus & Norma. The Wolf or Beast Centaurus' kill, the beast version being half man, half lion.
Lynx[28]	Fig. **92** NWsector *RA8h±,* *Decl45°±*	Lynx.
Lyra[52]	Fig. **89** NWsector *RA19h±,* *Decl35°±*	Between Cygnus & Hercules. Its brightest star Vega is one corner of the well-known asterism Summer Triangle. The Lyre of Orpheus Orpheus' music charmed animate and inanimate objects. He was able to drown out the Sirens singing to the Argonauts (to lure them to their deaths).
Mensa[75]	Fig. **89** SEsector *RA5h±,* *Decl-80°±*	Table.
Microscopium[66]	Fig. **89** SWsector *RA21h±,* *Decl-35°±*	Microscope.
Monoceros[35]	Fig. **92** SWsector *RA7h±,*	Unicorn.

Constellation	Fig., Loc.	Neighbours, The Story
	Decl0°±	
Musca[77]	Fig. **92** SEsector *RA13h±,* *Decl-70°±*	Fly (an insect).
Norma[74]	Fig. **92** SEsector *RA16h±,* *Decl-50°±*	Normal (a right angle).
Octans[50]	Fig. **89** SWsector *RA22h±,* *Decl-80°±*	Eighth of a Circle.
Ophiuchus[11]	Fig. **92** SEsector *RA17h±,* *Decl-5°±*	Between Hercules & Libra, the **CE** cutting right through its middle. It is Voyager 1's current background constellation. The Healer or Serpent-Bearer Ophiuchus grasping a snake (represented by the Serpens constellations)
Orion[26]	Fig. **89** NEsector *RA6h±,* *Decl0°±*	Between Canis Major & Taurus, the **CE** cutting right through its middle. Contains the well-known asterism Orion's Belt (of 3 stars, aka the Three Kings). Its 2nd brightest star Betelgeuse is one corner of the well-known asterism Winter Triangle, as well as being inside the Winter Hexagon asterism which all collectively make up 7 of the 21 stars of first-order ***Magnitude*** (and that they are part of the Milky Way band comes as no surprise)! Orion is one

Constellation	Fig., Loc.	Neighbours, The Story
		of only three constellations that has two 1st Order *Magnitude* stars. The Hunter Son of Roman god Neptune (Greek god Poseidon)
Pavo[44]	Fig. **89** SWsector *RA20h±, Decl-65°±*	Peacock. This is Voyager 2's current background constellation.
Pegasus[7]	Fig. **89** NWsector *RA23h±, Decl20°±*	Between Cygnus & Pisces. Contains the Great Square asterism. The Winged Horse Possessing magical powers.
Perseus[24]	Fig. **89** NEsector *RA3h±, Decl45°±*	Between Andromeda & Auriga. The Hero Slew Medusa. Rescued Andromeda whom he married.
Phoenix[37]	Fig. **89** SEsector *RA1h±, Decl-45°±*	The mythical Phoenix.
Pictor[59]	Fig. **89** SEsector *RA6h±, Decl-55°±*	Painter.
Pisces[14]	Fig. **89** NEsector *RA1h±, Decl10°±*	Between fellow Zodiacs, Aries & Aquarius, just above the *CE*, the Ecliptic cutting the *CE* near here. The Fishes Represent Aphrodite and Eros (Venus and Cupid to Romans), escaping a monster and leaping into the sea.
Piscis Austrinis[60]	Fig. **89**	Between Aquarius & Grus.

Constellation	Fig., Loc.	Neighbours, The Story
	SWsector *RA22h±,* *Decl-30°±*	The Southern Fish cf. Pisces (plural) to its NNW, which are two fishes, the children of this Piscis (singular).
Puppis[20]	Fig. **92** SWsector *RA8h±,* *Decl-35°±*	East of its related constellations Carina & Vela. The Poop Of the old constellation Argo Navis (the ship of Jason and the Argonauts, in search of the Golden Fleece) which was divided into three constellations.
Pyxis[65]	Fig. **92** SWsector *RA9h±,* *Decl-30°±*	Mariner's Compass cf. constellation Circinus (Draftsman's Compass).
Reticulum[82]	Fig. **89** SEsector *RA4h±,* *Decl-65°±*	Reticule.
Sagitta[86]	Fig. **89** NWsector *RA20h±,* *Decl20°±*	Between Aquila & Cygnus. The Arrow It represents the weapon of many myths.
Sagittarius[15]	Fig. **89** SWsector *RA19h±,* *Decl-30°±*	Between fellow Zodiacs, Capricornus & Scorpius, the galactic centre is located at the end of the Teapot asterism's spout. The (centaur) Archer His arrow is aimed at Antares, at Scorpus' heart, ready to kill it should it dare to attack Hercules. Note: Centaurus also represents a centaur albeit a different one.

Constellation	Fig., Loc.	Neighbours, <u>The Story</u>
Scorpius[33]	Fig. **92** SEsector *RA17h±, Decl-30°±*	Between fellow Zodiacs, Libra & Sagittarius. <u>The Scorpion</u> Goddesses sent it to battle Orion the Hunter who had vowed to hunt down all animals. It killed him!
Sculptor[36]	Fig. **89** SWsector *RA0h±, Decl-30°±*	Sculptor. Home of the Galactic South Pole.
Scutum[84]	Fig. **89** SWsector *RA19h±, Decl-10°±*	Shield.
Serpens[23]	Fig. **92** NEsector *RA16h±, Decl10°±* & SEsector *RA18h±, Decl-10°±*	In two parts: Serpens Caput (head), Serpens Cauda (tail), the bottom of the former and the top of the latter cross the **CE** close to *RA15h*. Ophiuchus lies between these two parts. <u>The Serpent</u> Ophiuchus' serpent.
Sextans[47]	Fig. **92** SWsector *RA10h±, Decl-5°±*	Sextant.
Taurus[17]	Fig. **89** NEsector *RA5h±, Decl15°±*	Between fellow Zodiacs, Aries & Gemini. The Galactic Plane's Anti-Centre 180° is near the tip of the Bull's northern horn. Its 1st Order *Magnitude* star Aldebaran is one corner of the Winter Hexagon asterism. <u>The Bull</u>

Constellation	Fig., Loc.	Neighbours, The Story
		This is Zeus in disguise when abducting Europa.
Telescopium[57]	Fig. **89** SWsector *RA19h±,* *Decl-40°±*	Telescope.
Triangulum Australe[83]	Fig. **92** SEsector *RA16h±,* *Decl-65°±*	Southern triangle, near equilateral.
Triangulum[78]	Fig. **89** NEsector *RA2h±,* *Decl35°±*	Triangle, almost isosceles.
Tucana[48]	Fig. **89** SWsector *RA0h±,* *Decl-65°±*	Toucan.
Ursa Major[3]	Fig. **92** NWsector *RA11h±,* *Decl55°±*	Between Bootes & Lynx, has the well-known Big Dipper/ladle asterism. <u>The Great Bear</u> Juno turns her husband Jupiter's beautiful mistress into this unattractive bear.
Ursa Minor[56]	Fig. **92** NEsector *RA15h±,* *Decl75°±*	Between Cassiopeia & Draco, also known as the Little Dipper/ladle. <u>The Lesser Bear</u> Although the smaller kin of Ursa Major, this constellation is more famous as it contains the North star, Polaris.
Vela[32]	Fig. **92** SWsector *RA9h±,*	West of its related constellations Carina & Puppis. Its stars Alsephina and Markeb are part of

Constellation	Fig., Loc.	Neighbours, The Story
	Decl-45°±	the False Cross asterism. Its star Suhail was given the name Regor (Roger spelt backwards) by Apollo 1 astronaut Gus Grissom as a joke on his fellow astronaut Roger Chaffee; tragically both, along with Ed White, perished in a fire in their capsule during a test in January 1967. [Guys, you made it into space, for there is a star named after one of your team. RIP.] The Galactic 270° lies in this constellation. The Sails Of the old constellation Argo Navis (the ship of Jason and the Argonauts, in search of the Golden Fleece) which was divided into three constellations.
Virgo²	Fig. **92** NEsector *RA13h±, Decl0°±*	Between fellow Zodiacs, Leo & Libra, the **CE** cutting right through its middle. Its brightest star is one corner of the asterism Spring Triangle, which in turn is the lower half of the Great Diamond asterism (see constellation Canes Venatici). The Virgin A virgin goddess.
Volans⁷⁶	Fig. **92** SWsector *RA8h±, Decl-70°±*	Flying fish.
Vulpecula⁵⁵	Fig. **89** NWsector	Little fox.

Constellation	Fig., Loc.	Neighbours, **The Story**
	RA20h±, Decl25°±	

Table of Asterisms, their Constellations

Asterism	Constellation
Big Dipper	Ursa Major
False Cross	Carina, Vela
Great Diamond	Bootes, Canes Venatici & Virgo
Great Square	Pegasus
Keystone	Hercules
Little Dipper	Ursa Minor
Northern Cross	Cygnus
Orion's Belt	Orion
Southern Cross	Crux
Spring Triangle	Bootes, Leo & Virgo
Summer Triangle	Aquila, Cygnus & Lyra
Teapot	Sagittarius
Winter Hexagon	Auriga, Canis Major, Canis Minor, Gemini, Orion & Taurus
Winter Triangle	Canis Major, Canis Minor & Betelgeuse

Printed in Great Britain
by Amazon

31977801R00158